# Instability, Skew-T, and Hodograph Handbook

Tim Vasquez

2017

Classical thermodynamics is the only physical theory of universal content that I am convinced, within the framework of its basic concepts, will never be overthrown.

ALBERT EINSTEIN, 1946
*Autobiographical Notes*

INSTABILITY, SKEW-T, AND HODOGRAPH HANDBOOK

Edition 1 | December 2017
Copyright ©2017 Tim Vasquez
All rights reserved

For information about permission to reproduce selections from this book, write to Weather Graphics Technologies, P.O. Box 450211, Garland TX 75045 or servicedesk@weathergraphics.com. No part of this publication may be reproduced, stored in a retrieval system, or transmitted by any means without the express written permission of the publisher.

ISBN 978-0-9969423-3-1
Printed in the United States of America

ISH A2017120500

# Contents

## 1 | Thermodynamics / 1
1.1. The parcel / 1
1.2. Altitude / 1
1.3. Pressure / 4
1.4. Physical units / 6
1.5. Heat / 8
1.6. Phases of matter / 15
1.7. Phase changes / 15
1.8. Temperature / 17
1.9. Water / 19
1.10. The saturation process / 21
1.11. Moisture properties / 25
1.12. Troposphere and stratosphere / 29
1.13. Scales / 30

## 2 | Observing systems / 35
2.1. History / 35
2.2. Balloons / 37
2.3. Pilot balloon / 38
2.4. Modern radiosondes / 39
2.5. Radar velocity data / 54
2.6. Wind profiler / 56
2.7. AMDAR / ACARS / MDCRS / 58

## 3 | Skew-T log-p diagram / 63
3.1. The skew-T log-p chart / 65
3.2. Construction of the skew-T / 65
3.3. Meteorological data / 68
3.4. Static stability and lapse rate / 70
3.5. Lifting / 77
3.6. Basic parcel properties / 80
3.7. Proximity sounding / 85
3.8. Parcel techniques / 86
3.9. Virtual temperature / 92
3.10. Dryline / 92
3.11. Capping inversion / 96

## 4 | Shear & hodograph / 103
4.1. The hodograph / 103
4.2. Shear / 107
4.3. Storm motion / 108
4.4. Storm-relative frame / 115
4.5. Shear and storm forecasting / 116
4.6. Helicity / 124

## APPENDIX / 131

### A-1 — Severe weather indices / 131
A1.2. Case study values / 131

### A-2 — SHARPpy / 185
A2.1. SHARPpy / 185
A2.2. Concept of operation / 186
A2.3 Sounding / 187
A2.4. Hodograph / 189
A2.5. Storm slinky / 190
A2.6. Equiv. potential temp. profile / 192
A2.7. Storm-relative wind profile / 192
A2.8. Possible Hazard Type / 193
A2.9. Thermodynamic output / 195
A2.10. Shear output / 199
A2.11. Sounding analogs (SARS) / 201
A2.12. Significant Tornado Parameter (STP) / 202

### A-3 — RAOB / 205
A3.1. Data import / 205
A3.2. RAOB major modules / 206
A3.5. Other features / 208

### A-4 — Other resources / 211
A4.1. University of Wyoming / 211
A4.2. UCAR weather / 212
A4.3. Storm Prediction Center / 213
A4.4. PivotalWeather.com / 214
A4.5. ESRL RUC Soundings site / 215

### A-5 — Radiosonde data / 217

### A-6 — Glossary / 221

## References / 225

## Index / 231

# INTRODUCTION

This book is written as both a reference and an introduction to instability and shear for the practicing forecaster, including emergency managers, broadcasters, and of course, government forecasters. One of the questions I get asked most commonly in e-mails, at conferences, and my online one-on-one training is how to work with soundings and hodographs. Incredibly enough, I have never covered this subject in depth until now.

The goal here has been to simplify the subject and make it accessible to the practicing forecaster. It's pared down to the essentials needed to understand instability and shear, with a steady progression of building onto the previous material. For a more complete physics and math picture, a search of booksellers will turn up many good textbooks on thermodynamics. What is in this book, however, is a bare minimum of what is sufficient for operational forecasting. I have omitted material that new forecasters are likely to skip. I had even considered eliminating SI units for rates using the old solidus form, e.g. using m/s instead of m s$^{-1}$ to express rates, for readability.

The Skew T diagram and hodograph, in fact, were developed to remove many of the complexities of atmospheric physics from the forecast process. Therefore you shouldn't let these tools discourage you. For advanced readers I have provided references to appropriate source material for more information or clarification.

This book does not cover storm structure, storm dynamics, or analysis. This is reviewed in my other severe weather book *Severe Storm Forecasting*. The focus here is on hodographs, soundings, shear, instability, and severe weather indices.

As with all my forecasting titles, I am committed to eliminating any errors that are found in future editions. You may send these to the address below and you will be credited when there is a reprint. I am deeply appreciative of your support over the years. This has been essential for allowing me to continue developing new forecaster titles and tools, and allowing me to provide the Meteorology Lab instruction project on YouTube.

Special thanks to Chuck Doswell, Mitchell Moncrieff, Jon Davies, and John M. Lewis; and with Alan Moller's passing I thank him for, many years ago, showing that there is an *art* to severe storm forecasting as well as a science.

TIM VASQUEZ
tim.vasquez@weathergraphics.com
December 1, 2017

**Table X-XX. Symbols for common variables used in thermodynamic operational meteorology**, with typical units.  Greek letters are provided in brackets for those not familiar with the letters.

### Latin alphabet

| | |
|---|---|
| $B$ | buoyancy; CAPE (J kg$^{-1}$) |
| $c$ | specific heat capacity (J kg K$^{-1}$) |
| $e$ | water vapor pressure (hPa, mb, mmHg) |
| $f$ | wind speed (m s$^{-1}$), frequency (Hz) |
| $g$ | gravitational constant (9.81 m s$^{-2}$) |
| $H$ | helicity (m$^2$ s$^{-2}$); height (ft) in barometry |
| $K$ | kinetic energy (J) |
| $m$ | mass (kg) |
| $p$ | pressure (hPa, in Hg) |
| $P$ | potential energy (J); precip. rate (mm h$^{-1}$) |
| $q$ | specific humidity (g kg$^{-1}$) |
| $r$ | mixing ratio (g kg$^{-1}$); reduction ratio |
| $R$ | gas constant; Earth's radius (~6371 km) |
| $S$ | shear magnitude (s$^{-1}$) |
| $t$ | time (s); Celsius temperature (rare) |
| $T$ | absolute temperature (°C, K) |
| $U$ | shear (m s$^{-1}$) |
| $w$ | mixing ratio (mass fraction) |
| $W$ | precipitable water (in) |
| $v$ | specific volume (m$^3$ kg$^{-1}$) |
| $V$ | volume (m$^3$) |
| $z$ | geometric height (m) |
| $Z$ | geopotential height (gpm) |

### Greek alphabet

| | |
|---|---|
| $\alpha$ | absorptivity [small alpha] |
| $\Gamma$ | lapse rate (°C km$^{-1}$) [lg gamma] |
| $\delta$ | differential (non-exact, geometric) [sm delta] |
| $\Delta$ | finite difference [lg delta] |
| $\varepsilon$ | emissivity [small epsilon] |
| $\theta$ | potential temperature [sm theta] |
| $\lambda$ | longitude, wavelength [sm lambda] |
| $\rho$ | density (kg m$^{-3}$) [sm rho] |
| $\Phi$ | geopotential [lg phi] |
| $\varphi$ | latitude [sm open phi, LaTeX :\varphi] |
| $\phi$ | latitude [sm closed phi, LaTeX:\phi] |
| $\omega$ | angular velocity (s) [sm omega] |

### Subscripted/superscripted variables

| | |
|---|---|
| $\zeta_i$ | mixing ratio (mass ratio) [sm zeta] |
| $T^*$ | virtual temperature (more commonly $T_v$) |

**Subscripts**, which modify a variable, with common examples. The use of a number in meteorology indicates a height or level (e.g., $T_{850}$). Large numbers usually indicate mb (hPa). Small numbers usually indicate km.

| | |
|---|---|
| 0 | sea level ($g_0$) |
| a | airfield elevation, ft ($H_a$); airfield pressure ($p_a$) |
| d | dewpoint ($T_d$); dry ($m_d$); dry adiabatic ($\Gamma_d$) |
| e | equivalent ($\theta_e$, $T_e$); environment ($\Gamma_e$) |
| i | ice ($e_i$) |
| ie | equivalent ($T_{ie}$) |
| m | moist; moist adiabatic ($\Gamma_m$) |
| p | constant pressure; station elevation, ft ($H_p$) |
| P | parcel ($T_p$) |
| s | surface; station pressure ($p_s$) |
| SR | storm relative |
| v | virtual ($T_v$, $\theta_v$); water vapor ($m_v$) |
| w | water ($e_w$); wet bulb ($T_w$, $\theta_w$) |
| z | sensor elevation, ft ($H_z$) |

# MATHEMATICAL NOTES

I decided with this book to cease use of the informal solidus system for division of units, e.g. m/s, cm/h, g/kg, etc, and begin writing these expressions as m s$^{-1}$, cm h$^{-1}$, g kg$^{-1}$, etc, as prescribed by SI (International System of Units) guidelines. I had used the solidus system for readability in all my earlier books, as it's more familiar to the layperson. My highest priority is making technical information accessible as possible, avoiding the use of formidable mathematics in introductory books for operational forecasting. Instruction through the mathematical approach is readily available in many other textbooks. Some is still necessary here, however.

Introductory science appears to be moving strongly towards SI notation. This was used for decades strictly in technical journals and university classrooms, but seems to have surged onto the Internet in everything from Powerpoint presentations to course notes. So it's time to get familiar with this system.

In SI, we use negative exponents for mathematical expressions indicating a division of units, which occurs anywhere "per" can be used. An example of this is *meters per second* (velocity). This is written as either m s$^{-1}$ or m·s$^{-1}$, both of which are the same. A negative exponent, as in $x^{-1}$, is equal to $1/x$. This divides instead of multiplies the term. So m s$^{-1}$ shows that meters are divided by seconds instead of multiplied, yielding meters per second. A higher exponent like m s$^{-2}$ indicates meters per second per second (acceleration), the rate of change of velocity. (If we want to go even further, there is m s$^{-3}$, which is the rate of change of acceleration, called jerk, and m s$^{-4}$, which is the rate of change of jerk, which is called jounce!)

Finally we often come across inverse seconds, s$^{-1}$. This means "per second", and is equal to cycles per second, or Hertz. It it is often used for shear, given in m s$^{-1}$ m$^{-1}$. Meters divided by meters cancels each other out, leaving the inverse second s$^{-1}$. Inverse seconds can also used for vorticity, expressing rotations per second.

Mastery of algebra and calculus is not necessary for understanding shear and instability as a forecaster, but for those who need clarification I recommend the YouTube channels Khan Academy, MathDoctorBob, and PatrickJMT. Simply visit the channel page and use the search box to find videos for the material you are interested in, such as negative exponents. If you get confused, drop down to a simpler level. Also there are excellent open courses offered by colleges which offer free access. To engage with the material even further, look for math textbooks to provide you with problems to work through, and use them in combination with the videos. When it comes to working through problems, there are no shortcuts. The books and lots of practice are necessary to thoroughly learn the material.

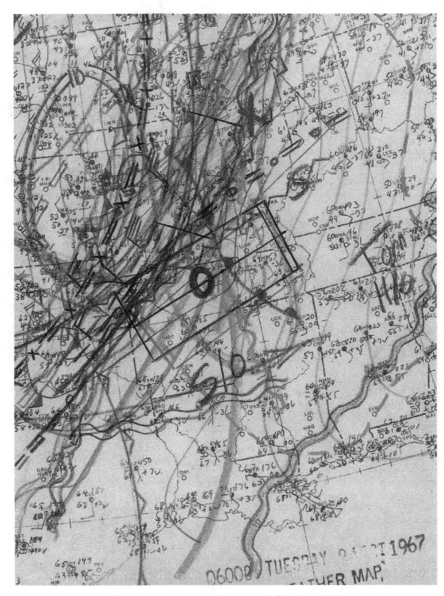

Plot for a severe weather case in October 1967
by severe weather forecasting techniques pioneer Col. Robert C. Miller.

Thanks to John M. Lewis and L. Wilson.

# 1 THERMODYNAMICS

# THERMODYNAMICS

Before diving into soundings and hodographs, forecasters need to be grounded in terminology, units of measurement, and physical processes. This constitutes the underlying physics behind the skew-T log-p diagram and explains much of the weather that forecasters encounter daily.

Heat is the source of all weather, and is also the basis for this book. Thermodynamics is defined as the relationship between heat and other forms of energy: electromagnetic, mechanical, and chemical energy are some of the ones we will cover. Convective heat transfer, for example, is what powers thunderstorms.

## 1.1. The parcel

Throughout this book we work with a *parcel*, which is a volume of arbitrary size that represents a small portion of the air mass and is tracked over time. The American Meteorological Society glossary defines a parcel as "an imaginary volume of fluid to which may be assigned various thermodynamic and kinematic quantities" (AMS 2017).

In thermodynamics, including work with the thermodynamic diagram, a parcel is frequently used to simulate processes occurring within a larger system. For example, a parcel can represent inflow and updraft air within a thunderstorm, helping to determine the forces that act upon it and how it will respond. In much of this book, a parcel can be thought of as having a scale of centimeters or meters in size.

## 1.2. Altitude

Before we begin working in vertical coordinates, we need to have a thorough understanding of height and altitude. Height, of course, expresses the vertical distance between two points. Altitude is the height above a given datum (typically mean sea level). We often assume heights in meteorology to be an expression of altitude above sea level, but there are actually several different types of altitude.

**Radiation smarts**
Is thermal radiation related somehow to uranium one might find in a nuclear power plant? There are actually different types of radiation, and it's important to understand the difference. Thermal radiation is transmitted through electromagnetism. The radiation we associate with "radioactivity" is particle radiation, in which a substance undergoes atomic decay and emits a particle. Particle radiation may be comprised of alpha particles, which can be stopped by a sheet of paper, and beta particles, which can be stopped by a piece of aluminum foil. Radioactivity may also involve gamma or X-ray radiation, which is not made up of particles but is a type of very high-frequency electromagnetic radiation at a shorter wavelength than visible light. Gamma radiation is particularly hazardous because it can break down DNA molecules.

## Pressure altitude

Pressure altitude is higher than station elevation in low pressure events, and lower than the station elevation in high pressure situations; this variation averages about 100 feet per 0.10 inch difference.

At jet cruising altitudes in the upper troposphere and stratosphere, by international agreement aircraft altimeter settings are set to mean sea level pressure (29.92 in Hg, 1013.25 mb) regardless of the actual pressure. Thus the altimeter displays pressure altitude, not geometric altitude. This is done to eliminate constant altitude changes due to the pressure varying along the route of flight.

### 1.2.1. Geometric altitude

Geometric altitude, $z$, also known as true altitude, is the vertical distance measured from mean sea level (MSL) to the given level, as might be measured with a ruler. Elevation referencences the true altitude of the Earth's surface.

### 1.2.2. Geopotential

Geopotential, $\Phi$, is an expression of the work required to raise a mass from sea level to a given altitude. This is a fairly simple concept: $\Phi = g \Delta z$, gravity multiplied by the change in height. At sea level, it equals zero.

Since $g$ is m·s$^{-1}$ and $z$ is m, we get m·s$^{-2}$·m, which combines to m$^2$·s$^{-2}$. This is a unit of energy, because joules, a unit of energy, are expressed as kg·m$^2$·s$^{-2}$. So substituing we get J=kg·$\Phi$. Rearranging this we get $\Phi$=J·kg$^{-1}$, showing that geopotential equals energy per kilogram of mass.

### 1.2.3. Surface gravity

Earth has an oblate shape, caused by centrifugal force of the rotating planet acting on the lithosphere over billions of years. The absolute altitude (distance from the center of the earth) of sea level is 21 km higher at the equator than it is at the poles. Since gravity weakens with distance from the Earth's center, surface gravity is 0.2% weaker at the Equator than at the poles. Centrifugal force, or centripetal acceleration, from the Earth's spin decreases gravity at the Equator by 0.3%.

Taking everything into account, we get values of $g$=9.832 m s$^{-2}$ for the poles and $g$=9.780 m s$^{-2}$ for the equator. This is a difference of about 0.5%.

### 1.2.4. Geopotential height

Geopotential height, $Z$, also known as dynamic height, is given by the equation $Z = \Phi/g_0$, or $Z = g\Delta z/g_0$, where $Z$ is geopotential height in meters, $\Phi$ is geopotential and $g_0$ is standard gravity, which equals 9.807 m s$^{-2}$.

Consider the concept of geopotential "surfaces", where geopotential is constant. On such a surface, all objects of similar mass have the same potential energy, and cannot change to a different surface without gaining or losing

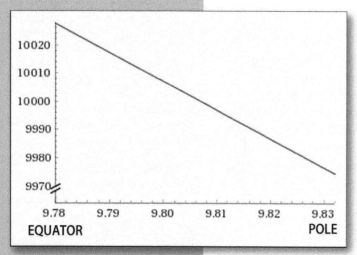

**Figure 1-1. Geopotential height** with respect to gravity and latitude, for a height of Z=10,000 m. Geopotential is about 0.5% lower at the poles.

potential energy. We can rewrite the equation to solve for geometric height: $\Delta z = Z \cdot g_0/g$. Without solving anything, we can see that with gravity $g$ being stronger in the polar regions, the equation's denominator will be larger, and so geopotential will be smaller in the polar regions.

Let's use a constant geopotential height of 10,000 gpm. Using gravity over the poles, we get $\Delta z$=9975 m, and over the equator we get $\Delta z$=10028 m. So this shows that the geopotential surfaces are closer to the poles than they are over the equatorial regions.

Most of our references to height are actually geopotential height, $Z$, where formulas use $g_0$ (standard gravity) instead of $g$ for gravity. The familiar 500 mb chart shows contours in geopotential decameters. However our example above shows that the variation is minor for a 10 km layer, about 0.2%, so for general descriptions of a forecast environment we can assume $z$ and $Z$ are equivalent.

### 1.2.5. Absolute altitude

The altitude with respect to the center of the earth is known as the absolute altitude. In aviation it is the geometric height above ground level (AGL) and can be measured with a radar altimeter.

### 1.2.6. Pressure altitude (aviation)

Pressure altitude (QNE in aviation) provides the geometric altitude above the pressure surface which equals mean sea level pressure (29.92 in Hg or 1013.25 mb). When an altimeter is set to 29.92 it will display the QNE value, the pressure altitude. This is used in aviation but not in operational forecasting.

### 1.2.7. Indicated altitude (aviation)

Indicated altitude is the altitude that shows on the altimeter when it is set to the correct sea-level pressure. This is not used in operational forecasting.

### 1.2.8. Density altitude

Density altitude is the pressure altitude corrected for deviation from standard temperature. This is important for aviation but is not used in operational forecasting.

---

**Wind & pressure theories, 1720s**

The clouds are high, because the air on which they are suspended is heavy. For when the air near the Earth is more dense than usual, then it becomes heavier than the vapours which must then ascend, and at least settle in the regions of the air, which is of the same gravity with themselves.

If the vapours were the cause of the increase of the air's gravity, there must be as many vapours in the air at a time, as are equal in weight to three inches of mercury. For so much we find the mercury rises or falls in the weatherglass. Now mercury is about 14 times heavier than water, and consequently there must be in the air at once as many vapours as are equal to a column of water of 42 inches, and whose base is equal to the surface of the earth. And [J. Ozanam] concludes that the reason why air is heavier at one time than another arises more probably from there being more air on that part of the earthly surface when the air grows heavier, and this proceeds from winds.

For example if the wind (which as the Doctor says) is nothing else but a stream of air, should blow on any place, and the air thus moved should be pent in that place by mountains or hills: or if two contrary winds should blow on the same place, in either case the air will be heaped up in the middle. Consequently there being more air its gravity will increase. But if the aforesaid circumstances or the like don't happen in any particular country, then the air over which it grows less in quantity and consequently lighter. Whence it is plain that winds are the only causes of the variation of the air's gravity.

JOHN T. DESAGUILIERS, c. 1720
Natural philosopher & engineer
*Notes on Ozanam's Hydrostatics*

## THERMODYNAMICS

**Q-codes**
The Q-codes for pressure originate from radio brevity codes governed by the International Telecommunication Union and designed in the 1940s for both morse code and voice. Most are no longer used, however about 15 codes still remain, including the barometric ones.

QFE - Query Field Elevation
QFF - Query (meteorological office)
QNH - Query Nautical Height
QNE - Query Nautical Equivalent

**Explaining steam**
A common misconception is that steam is the visible cloud that comes out of a locomotive or kettle. This visible plume is a condensation cloud actually made up of liquid water droplets, the result of steam condensing in the cooler surroundings. Steam itself, like water vapor, is invisible.

**Mean sea level**
Sea level itself is not flat. Satellite measurements show the tropical Pacific is 1 m above MSL while while the polar oceans are about 1 m below MSL. The maximum sea height is found along the eastern Japanese coast and the lowest along the Antarctic coast.

## 1.3. Pressure

Pressure, more specifically atmospheric pressure, is an expression of the force acting on a specific area. It is measured in mb in US forecasting, and hectopascals (hPa) worldwide. Both are equal. It may also be measured in inches of mercury (in Hg), which equals hPa / 33.8636.

Barometric pressure is often an ambiguous term in meteorology, but often refers to altimeter setting when in inches, and sea level pressure when in millibars. It normally involves a "reduction" to sea level, in order to standardize barometers throughout a forecast area at many different elevations to a common level.

### 1.3.1. Pressure

Pressure, $p$, $p_s$, and $p_{sta}$, refers to *station pressure*, or *sensor pressure*, (QFE in aviation) which is the true pressure of the atmosphere without considering any reduction or correction to sea level. This is the pressure as would be measured directly by a calibrated barometer transported from the factory to the measurement location, at a different elevation without corrections. It equals 0 in the vacuum of space and 29.92 in Hg in standard conditions at MSL. Normally mb (hPa) are used.

Weather station aneroid and mercury barometers always read an uncorrected value, indicating the station pressure. The readings must be reduced to sea level using a designated formula, a reference table, or a computer program. Station pressure is never transmitted in METAR weather reports, but is carried in SYNOP code.

### 1.3.2. Altimeter setting

Altimeter setting, $p_{alt}$, (ALSTG, QNH in aviation) is an expression of sea level pressure, as station pressure reduced to MSL via the ISA standard atmosphere, a model for typical atmospheric conditions. This makes it useful for providing data to aircraft so that their instruments read true altitude. The equation is (NWS 1998):

$$p_{alt} = (p_s^{0.190263} + (1.313 \cdot 10^{-5}) \cdot H_a)^{5.25588}$$

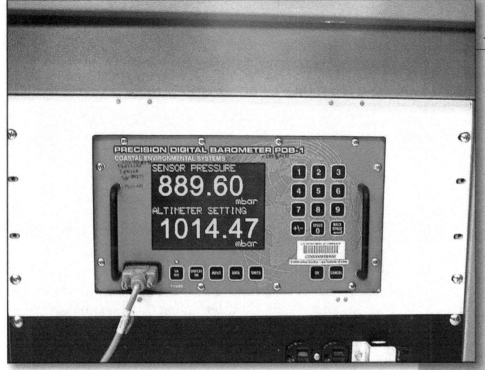

**Figure 1-2. Precision barometer** used in a National Weather Service office. This reads sensor pressure, $p_s$, and altimeter setting, $p_{alt}$, which is a simple reduction to sea level. (NOAA)

where $p_{alt}$ is the altimeter setting, $p_s$ is the station pressure, and $H_a$ is the airfield altitude. This shows that the only variables needed are station pressure and height.

This value is passed to the public as the barometric pressure for the city, since it is expressed using inches of mercury (in Hg), the same units printed on all consumer barometers. Altimeter setting is transmitted in METAR but not in SYNOP reports.

If an aircraft altimeter is set to this value, the instrument will show the true altitude. When a weather map or bulletin shows pressure in inches of mercury, it typically refers to altimeter setting.

### 1.3.3. Sea level pressure

Sea level pressure, $p_{sl}$, (SLP, QFF in aviation) is the barometric pressure corrected to sea level, which attempts to compensate for the effects of diurnal atmospheric tides and abnormal atmospheric density. In the United States it is calculated as (NWS 1998):

$$p_{slp} = (p_s \cdot r + C)$$

where $p_{slp}$ is sea level pressure, $p_s$ is the station pressure, $r$ is a reduction coefficient supplied by the National Weather Service, and $C$ is a correction factor not given here, which

# 6 THERMODYNAMICS

Table 1-1. **Pressure conversation chart**. U.S. Standard Atmosphere (1976) values of geopotential altitude and temperature are listed below.

| hPa/mb | 1013.25 | 1000 | 925 | 850 | 700 | 500 | 400 | 300 | 250 | 200 | 100 |
|---|---|---|---|---|---|---|---|---|---|---|---|
| bar | 1.0133 | 1 | 0.925 | 0.85 | 0.7 | 0.5 | 0.4 | 0.3 | 0.25 | 0.2 | 0.1 |
| in Hg | 29.92 | 29.530 | 27.315 | 25.101 | 20.672 | 14.765 | 11.812 | 8.859 | 7.382 | 5.906 | 2.953 |
| Pa | 101325 | 100000 | 92500 | 85000 | 70000 | 50000 | 40000 | 30000 | 25000 | 20000 | 10000 |
| psi | 14.670 | 14.503 | 13.416 | 12.328 | 10.153 | 7.2519 | 5.8015 | 4.3511 | 3.6259 | 2.9008 | 1.4504 |
| atm | 1 | 0.9869 | 0.9129 | 0.8389 | 0.6908 | 0.4935 | 0.3948 | 0.2961 | 0.2467 | 0.1974 | 0.0987 |
| torr | 760 | 750.06 | 693.81 | 637.55 | 525.04 | 375.03 | 300.02 | 225.02 | 187.52 | 150.01 | 75.006 |
| Geopotential alt. (m) | 0 | 111 | 762 | 1457 | 3012 | 5574 | 7185 | 9164 | 10363 | 11784 | 16180 |
| Geopotential alt. (ft) | 0 | 364 | 2500 | 4781 | 9883 | 18289 | 23574 | 30066 | 33999 | 38661 | 53083 |
| Temperature (°C) | 15.0 | 14.3 | 10.0 | 5.5 | -4.6 | -21.2 | -31.7 | -44.6 | -52.4 | -56.5 | -56.5 |
| Temperature (K) | 288.2 | 287.4 | 283.2 | 278.7 | 268.6 | 251.9 | 241.4 | 228.6 | 220.8 | 216.7 | 216.7 |
| Potential temp. (K) | 287.1 | 287.4 | 289.5 | 291.9 | 297.4 | 307.2 | 313.8 | 322.5 | 328.2 | 343.3 | 418.6 |

requires virtual temperature, station pressure, and station elevation. As with the altimeter setting equation, the structure is more important for our review than the actual numbers.

This makes it more useful for comparing the station to other stations in the region and constructing isobars. Each country prescribes its own method of calculating it. In the United States, a mean 24-hour temperature and station pressure are determined, and a reduction coefficient is obtained from tables prepared by a barometry specialist at the National Climate Data Center or National Weather Service headquarters. When a weather map or bulletin shows pressure in millibars, it typically refers to sea-level pressure. This is transmitted in both SYNOP and METAR reports.

## 1.4. Physical units

### 1.4.1. Temperature

Temperature, $T$, is an expression of the average molecular kinetic energy in a system. This system can be a raindrop, a parcel of air, or the air mass over Wyoming. Most of us are familiar with degrees Celsius °C, or degrees Fahrenheit, °F, and in meteorology we do use them for various expressions. However it's important to measure physical processes in Kelvin, K, which are not degrees but their own units, i.e. 273 K, not 273 °K.

Kelvin is exactly the same as Celsius, except the base is different: 0 °C represents the melting point of water, while 0 K represents absolute zero, the lowest possible temperature representing no molecular energy. In forecasting we fre-

---

**Total solar irradiance**
The amount of heat reaching the Earth every second is 100 quadrillion. This is the same amount of energy that was released by the 2004 Indian Ocean earthquake, as well as the 50 megaton Tsar Bomba that was detonated in 1961.

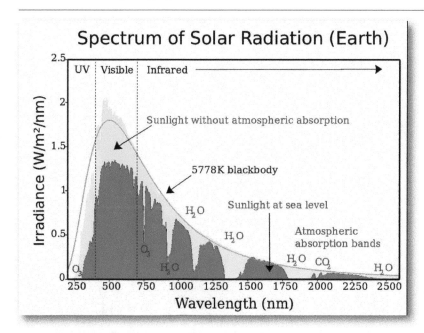

Figure 1-3. Electromagnetic radiation spectrum of the sun, ranging from high-frequency short wavelength emissions on the left, and low-frequency long wave emissions on the right. The top yellow line shows the radiation received at the top of the atmosphere, with the bottom red line showing what is received at the bottom. Large differences between the two at a given x-axis position indicate attenuation by layers that interfere with the incoming radiation at that wavelength or frequency. A good example is the ozone layer, which absorbs ultraviolet light at the very left side of the diagram. Attenuation in the infrared bands is due to water vapor. Outgoing longwave radiation of Earth is at 4000 to 100,000 nm, off the chart's right side. Incoming shortwave radiation is all of the energy shown here. *(Solar spectrum en.svg, from Wikipedia, by Nick84, Creative Commons)*

quently convert Kelvin to Celsius, so memorizing the relationship 0 K = 273.15 °C is important.

### 1.4.2. Mass

Mass, $m$, is a property describing the amount of matter in an object or parcel. It is a function of the number of molecules multiplied by the atomic mass of the molecules, so hydrogen 1 amu has much smaller mass than nitrogen, which has 14 amu. This becomes important when we discuss the mass of dry air, without water vapor, and moist air, which contains a certain amount of water vapor comprised of different molecules.

### 1.4.3. Volume

Volume, $V$, describes the size of a mass in three dimensions. It is expressed in m$^{-3}$, cubic meters. When talking about the physics of the atmosphere we often look at relationships with arbitrary volumes, so when we look at the relationship $V = 1/\rho$, we can see that when density, $\rho$, increases, the volume must decrease.

### 1.4.4. Density

Density is a property made up of mass per unit volume, $m/v$, and is measured in units of kg m$^{-3}$. Density is represented by the Greek letter rho, $\rho$, which looks similar to $p$, pressure, but is not the same thing, so readers need to be

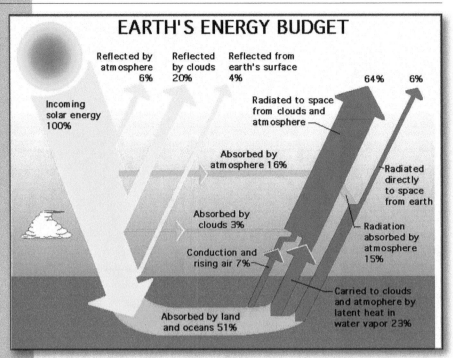

**Figure 1-4. The Earth's radiation budget.** About 341 W m⁻² of incoming radiation enters the atmosphere. About 30% of it is reflected back out immediately, especially from masses with high albedo, such as snow cover, cumulus clouds, stratus layers, and dry desert. This leaves an average of 239 W m⁻² which reaches the ground to do work with land, sea, and air masses. Note that a massive amount of heat is stored in the land and sea, but it ultimately finds its way back into the atmosphere, with half of this producing weather and the other half being radiated out as long wave energy. *(Dr. Lin Chambers / NASA)*

careful. Air density has a strong relationship to temperature and to a lesser extent to moisture. It is also inversely related to volume, $V = 1/\rho$, which shows that when density becomes higher, the molecules concentrate into a much smaller volume, so V decreases.

## 1.5. Heat

Meteorologists deal with all forms of thermal energy transfer daily, often without thinking about it. Calculating this transfer is a tremendous challenge for numerical modeling teams, where some specialists deal with only heat, due to its complexity. The processes we see range from conduction over warm ocean currents, to long wave radiation from cities at night, to the insulative properties of fresh snow cover. The one we deal with most in this book is convection, the physical movement of heat, typically in the form of a cumulus or cumulonimbus cloud. However other processes like radiation and conduction are important in day-to-day forecasting.

There are four types of heat transfer: radiation which is transferred through electromagnetic waves; convection, which is transferred upward through buoyancy; advection, where external forces move the heat; and conduction which is transferred through direct contact.

Because these are the only types of heat transfer available, this makes air a good insulator. The molecules are spaced far apart. A hot object on one side of the insulator emits black body radiation which can transfer heat through the air, which is not very efficient.

It can also transfer heat through convective air movements, but this is not efficient either since the widely spaced air molecules cannot carry much of the heat. Finally the object can heat the air through conduction, however this is not particularly efficient either, again due to the spacing of the molecules. The insulative properties of air make the air feel stiflying on a hot day with no wind, as the heat is not carried away from the skin, while more heat continues to be generated by the body. The wind chill factor is an example of an index that considers the ability of wind to mechanically remove the heat from the body.

### 1.5.1. Radiation

The cycle of weather starts with the massive influx of energy from the sun. This type of energy is called radiation, consisting of electromagnetic energy. This is caused by vibration of molecules at a specific rates, which produces an electromagnetic wave, much like waves traveling through the ocean.

By measuring the time interval between a wave we get frequency. In electromagnetism, this defines what part of the spectrum the energy falls within. The waves travel at the speed of light, giving us a velocity, so we can convert frequency to wavelength, showing how far apart the waves are.

Wavelength and frequency are *inversely related*, as given in the formula is $\lambda = v / f$. A good way to think of it is watching waves rolling in on a beach. If we're watching 10 waves per minute arriving, and it changes to 5 waves per minute, we naturally assume that the velocity of the incoming waves has slowed or the interval between the waves, wavelength, has gotten larger.

Fortunately, electromagnetic energy travels at a fixed speed: $c$, the speed of light, which is a constant 300 million m s$^{-1}$. So if we're listening to a Taylor Swift song on 101.7 MHz, we get a wavelength of 300,000,000 / 101,700,000 = 2.95 m. The sun doesn't broadcast at a specific frequency, though; it emits radiation at all frequencies and all wavelengths, including 101.7 MHz, which we would hear on the radio as

**Figure 1-5. Arizona desert near Ajo.** This is a region of tremendous heat exchange, with large amounts of both incoming shortwave energy during the day and outgoing longwave energy during both the day and night. *(Author photo)*

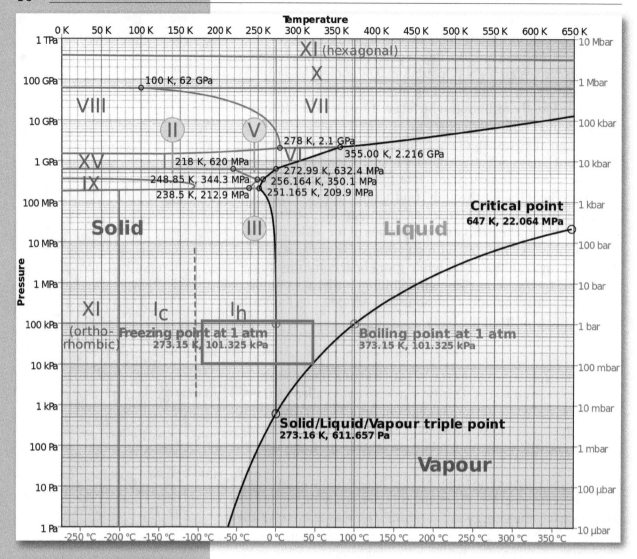

**Figure 1-6. Phase diagram for water.** This shows the phase of water which will occur at a given combination of temperature and pressure. Elevation decreases from bottom to top, so the bottom part of the diagram represents conditions close to outer space, while the red line at 100 kPa (1 bar; 1000 mb; 1000 hPa) represents conditions at sea level. The red box near the middle of the diagram represents conditions we deal with in operational meteorology. Conditions higher on the diagram are relevant to conditions at oceanic depths. The deepest part of the ocean is near the line at 1 kbar. If the oceans were deeper than about 60 miles, we would see unusual forms of ice (Ice VI, Ice VII, etc) due to the extremely tight packing of molecules. *(Cmglee, Creative Commons)*

**Figure 1-7. Planck's Law curve** for 313 K. This shows that most energy of a warm 104 °F sidewalk is concentrated at about 10 microns or 10,000 nanometers. This places it firmly in the far infrared or longwave infrared part of the spectrum. If it were thousands of degrees it would peak in the visible spectrum and we would see the sidewalk glow; it would likely be molten!.

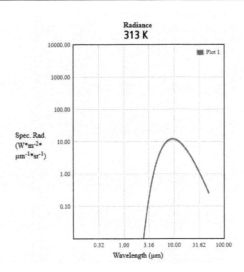

noise. However the energy levels on the radio bands are relatively low. The majority of the radiation arrives in the visible light band, which is an extremely high, very small wavelength, about 600 nanometers or 0.6 microns, and a huge amount also arrives in the infrared bands, which has a lower frequency and a larger wavelength. Radiation also arrives in the ultraviolet, X-ray, and gamma bands, but this has a negligable effect on weather.

In meteorology when dealing with temperature changes and the behavior of air masses, we deal with two main types of radiation: shortwave infrared radiation and its counterpart, longwave infrared radiation. The shortwave energy is close to the visible spectrum and it contains more energy than the long wave energy.

The Earth's surface, when the sun is at the zenith, receives on average 341 W m$^{-2}$ of irradiance, energy that can be converted to heat. The maximum possible is 1120 W m$^{-2}$. Most of the radiation arrives into the earth's atmosphere as shortwave infrared radiation, which heats the Earth's surface.

Electromagnetic radiation of the ground is a process called *insolation*, where the radiation from sunlight heats up dirt, rocks, pavement, vegetation, water, buildings, and other exposed masses. It is a misconception that the solar radiation heats up the air. There is some interaction with air molecules, but they are so far apart that there is little heating of the air from radiation. All of these processes are the same as what happens in a microwave oven: the food gets hot and the air does not heat up noticeably.

**Radiation observations, 1820s**

It has often struck me with surprise that in the numerous meteorological registers which are published in different parts of the world, no one has ever thought of including observations upon the intensity of the solar rays at different seasons of the year, and in different situations. It is well known to the agriculturalist and the gardener, whatever may be the temperature of the air, the fruits of the earth seldom come to perfection.

What, then, is the force of this important agent? what the modifications to which it is subject? and how is its energy spent when it is screened by concrete vapours from the surface of the earth? Does its influence increase with the temperature of the air from the pole to the equator? or is the rapid vegetation of the arctic regions, during the short summer of those climates, dependant on any compensating energy of its operation?

. . .

From these facts I conclude that the power of solar radiation in the atmosphere increases from the equator to the poles, and from below, upwards. The obstruction which the air offers to the passage of the rays is not alone dependant on its density at the surface of the earth, for most of the experiments which establish the difference between the lower and higher latitudes were made under nearly equal heights of the barometer.

JOHN FREDERIC DANIELL, 1827
*Meteorological Essays and Observations*

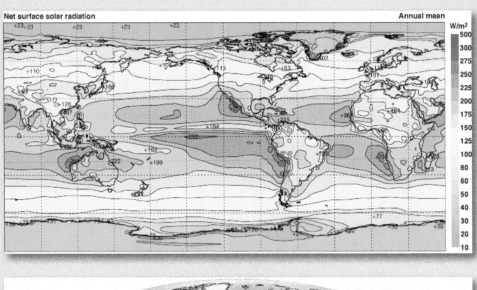

**Figure 1-8a. Incoming total radiation** map for the world. Without an atmosphere or weather, this map would take on a plain horizontally banded appearance. The variations are caused by clouds which intercept up to a quarter of the radiation and reflect it away, while up to another quarter is absorbed by the atmosphere, including by ozone. Note the minima near the equator, caused by the presence of the intertropical convergenence zone (ITCZ). *(Courtesy ECMWF)*

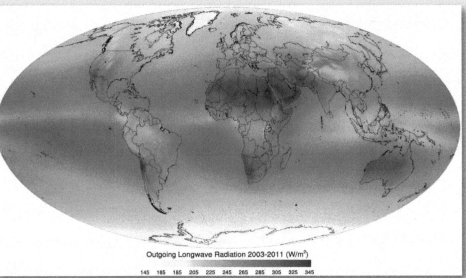

**Figure 1-8b. Outgoing longwave radiation** map for the world. The maximum radiation is emitted from the deserts. This is mostly due to the tremendous thermal energy which is radiated away at night under clear skies. (NASA

---

**Computing wet bulb**
Given a parcel temperature and dewpoint, the wet bulb temperature can be determined on the Skew T diagram by lifting the parcel from T, Td to the LCL, then descending along a moist adiabat to the original level. This is called a net isobaric process. The resulting wet bulb temperature is a close approximation to the actual wet bulb temperature that is found isobarically.

Once objects have heat, they naturally emit black body radiation. The electromagnetic wavelength is a function of temperature, but at the relatively low temperatures of 30°C, or 303 K, Planck's Law shows that radiation is concentrated in the longwave part of the spectrum. During the day the incoming short-wave radiation overwhelms the energy loss through longwave radiation, but at night, the longwave heat loss is quite significant. It is known as outgoing longwave radiation (OLR). It can be stopped by a layer of clouds, which reflects the radiation back to earth, and the air temperature slows or stops falling as a result.

Most places on earth radiate about 240 W m-2 of energy into space, with the deserts of Arizona and California having the strongest annual radiance in the United States. So about 20% of the incoming solar energy during the day is lost to space on a clear night, and the rest of the captured heat is available to warm the earth and create weather.

### 1.5.2. Convection

Convection is the vertical transfer, or diffusion of heat. We deal with *natural convection,* in meteorology, which arises due to buoyancy. The buoyancy comes from a lower density, which in turn is due to a higher temperature. Convection is what we see when a cumulus cloud is growing.

A warm mass of any fluid, such as air, water, or molasses, has low density compared to cooler portions of that fluid. In the presence of a gravitational field, low density produces high buoyancy.

The forces acting on a parcel of air are comprised of a downward vector made up of $\rho \cdot g$, density and gravity, while an upward vector $\Delta p/\Delta z$ represents the pressure gradient in the vertical, that is, the change in pressure per distance of geopotential height. If density is weaker, the downward vector diminishes, the upward vector dominates, and the parcel rises.

In the atmosphere we see this convection as rising thermals or clouds. By contrast, cold air is less buoyant and it tends to sink. The rising clouds and thermals leave a void underneath, and air moves with a horizontal component to fill this void. All of this motion produces a cell, where cool air sinks and warm air rises.

### 1.5.3. Advection

Advection is the forced movement of heat. An excellent example is wind, which perhaps moves warm air over Louisiana northward to Illinois, or movement of the air by a desk fan. Energy in the system is conserved, as the processes to move the parcel are expended by other processes.

Advection is not the same thing as convection. In meteorology, we consider the motion to be caused by outside processes like pressure gradient or terrain effects. However convection and advection often occur together naturally in unstable, sheared environments.

### 1.5.4. Conduction

Thermal energy may be conducted from one system to another, as in the case of the Earth's surface transferring heat

---

**Clausius-Clapeyron equation**

Mixing ratio is given by the following formula:

$$r = w = \frac{\varepsilon \cdot e}{P - e}$$

The quantity $\varepsilon$ is the ratio between the gas constant of dry air and the gas constant of water vapor 0.622 g.

The Clausius-Clapeyron equation expresses saturation vapor pressure as follows:

$$e_s = e_0 \cdot \exp\left[\frac{L}{R_v} \cdot \left(\frac{1}{T_0} - \frac{1}{T}\right)\right]$$

This shows that there is an exponential relationship between temperature and vapor pressure. This can be a significant factor in severe weather events.

**Figure 1-9. Joseph Black**, who in 1762 proposed latent heat.

**Joseph Black and latent heat**

The discovery of latent heat is attributed to the Scottish scientist Joseph Black. His students were sons of whiskey distillers, who met with Black and awarded him a commission to analyze the energy consumption of the distillation process in order to reduce the rising cost of fuel. Black proposed in 1762 that heat was absorbed when mash was boiled, slowing the heating process, while it was released when the mash was condensed. He also noted similar changes with the freezing and melting process.

In his lecture hall at the University of Glasgow he demonstrated the principle to his students with a block of ice and a container of water of equal mass, both measured to be 0 °C. He showed that over the course of up to an hour that the water warmed significantly, while the ice was sluggish to heat up. This, he showed, was caused by absorption and release of latent heat.

Black taught the principles of specific heat and discovered carbon dioxide. He was said to be a talented educator, attracting students and scientifically minded individuals from all over Britain. Unauthorized prints of his lectures, an early instance of pirating, were printed in 1770.

His contributions to latent heat did little to help the whiskey producers, but Black's close association with student James Watt lay the groundwork for the steam engine, and ultimately, for the Industrial Revolution.

to the atmosphere. Conduction occurs when the molecules from each system come into contact. The energy in the hot system are in close thermal contact with the other system and excite them, imparting energy to them at their own expense. If the systems remain in contact, this transfer of energy continues until the temperatures equalize. When this happens, thermal equilibrium has been reached. If the ground is warm enough, the conduction may immediately change over to convection, rising as thermals.

Another form of conduction occurs when a cold air mass moves over warmer ground or ocean waters. The bottom of the cold layer is heated, and the conducted heat immediately changes over to buoyant convection. This often causes gusty winds, extensive low-level turbulence, and sometimes cumuliform clouds. This type of conduction often occurs during the transition seasons, particularly in fall, in the wake behind Alberta clippers as polar high pressure advances southeastward with cold air and a strong pressure gradient.

There can also be a downward transfer of heat. This occurs when warm air advances over cold surface or ocean waters. If the air mass has a high humidity, the loss of heat to the ground may cool the air mass sufficiently to raise the humidity to its saturation point. Long-wave radiation of a landmass at night may increase the cooling even further. Fog and clouds then form. This is a common forecast problem along the Gulf Coast states during the cool season.

### 1.5.5. Phase changes

Heat can also be transferred through phase changes; that is, the change between gas, solid, and liquid. This involves the transfer of latent heat, which is used to break molecular bonds, but is given up when the bonds are reassembled. For example when water freezes, changing a liquid to a solid, the bonds are reassembled and latent heat is released into the system. This can delay the onset of the freezing process.

## 1.6. Phases of matter

There are three primary phases of matter that we refer to in meteorology: solids, liquids, and gases. Each of them have specific properties that are different from the others. Each phase has unique density, refraction index, and basic physical properties.

### 1.6.1. Solid
The solid phase of matter is structurally rigid and resists changes to its shape. In meteorology, the solid form of water is known as ice, and may take on other forms such as snow, freezing rain, glaze, graupel, and hail. The molecules are tightly packed together and are in a low energy state.

### 1.6.2. Liquid
The liquid phase of matter has a fixed volume but can change shape. In meteorology, the solid form of water is known simply as water, and may take on the form of rain, drizzle, or cloud droplets depending on the particle size. The molecules are bound together but able to move around.

### 1.6.3. Gas
The gaseous or vapor phase has a variable volume and variable shape. In meteorology, the gaseous form of water is always known as water vapor. Steam is a form of water vapor produced by sources of intense heat, but most of the water vapor condenses quickly into droplets once away from its immediate source.

## 1.7. Phase changes

The phase in which we find water is entirely dependent on the pressure and temperature. At sea level (100 kPa, 1 bar), for example, water will freeze below 273 K (0 °C), and will become vapor above 373 K (100 °C). We are extremely fortunate that within the troposphere, the freezing point is

**Origin of haze**
In the mid-20th century haze was believed to originate from tropical air masses. However field experiments found that visibility often improved with the arrival of the sea breeze. Computer trajectory analysis showed haze events in the United States were associated with stagnant polar air with long air mass loiter times over industrialized areas of the northeast US. However haze is now understood to not be a natural phenomenon.

In the US, studies showed a gradual decrease from the 1950s to the 1980s in mean visibilities east of the Rockies. Later studies have shown visibilities improving from the 1980s to present, probably due to stringent emissions standards, though the rise in major wildfires continues to cause haze trouble. Surges of Saharan dust embedded in deep tropical flow occasionally cause natural haze events in the southern US in midsummer.

In India, haze events in October and November are linked to the burning of leftover straw to rapidly clear fields for winter wheat. In Bangladesh, which has a major brick works industry, thousands of kilns burn 4 million tons of coal, firewood, rubber tires, recycled motor oil, trash, and plastics every year during the winter and spring before the monsoon rains.

China's notorious haze is caused by combustion from poorly-regulated factories, extensive use of coal as fuel, and millions of vehicle engines without the US EPA's aggressive standards. In China's stagnant winter weather patterns, the haze gets severe enough to go viral on social media. In 2007 China's ministry of health cited cancer from industrial pollution as the leading cause of death in China.

## 16   THERMODYNAMICS

---

**The barometer**

The baroscope, an instrument of a modern invention, whereby the authors pretend to discover the temper and inclination of the air; which must be of great use to those that are employed in gardening, agriculture, etc.

Now this column of quicksilver in the tube they do allege is supported by the weight of the ambient air, pressing on the stagnant quicksilver in the vessel, and that as the air becomes more or less ponderous, so does the quicksilver in the tube rise or fall more or less accordingly, but then in case the stagnant quicksilver were broader, in a broader vessel, the greater quantity of air would press harder upon it, and the quicksilver in the tube rise higher, but it does not.

If the mercury rises very high, the weather will continue fair so long as it stands at that pitch; and you will not find it change much, till the mercury falls down a good pace lower. So likewise when it is fallen down very low you must expect wet weather all the time of its so continuing, in both which particulars you will be certain, provided the wind and moom concur, for both the wind and the changes of the moon are to be observed in order to make a right prediction.

The mercury is always observed to be lowest in extreme high and strong winds, if it happens when the air is full of moisture, but the glass does no way predict winds beforehand, for the extreme lowest of the quicksilver only happens at the very time the wind blows, and as soon as the wind ceases, the mercury is then found to rise apace, but such a rise that immediately follows storms is no sign of fair weather, except it rises much higher than it was at the time of the wind's beginning to blow.

M. CHOMEL, 1725
Dictionaire Economique

---

constant and the vapor point does not become a factor; at the top of the troposphere the vapor point is 45 ˜C, far above the temperatures normally encountered there.

When there is a change in pressure or temperature, water may be forced to change phases, such as from a liquid to a gas.

### 1.7.1. Endothermic phase changes

When matter changes from a lower to a higher energy state, it is an endothermic process. Heat must be removed from the environment to carry out the phase change. This heat is "stored" within the water as latent heat. The cooling of the environment is very important to forecasters and it is one way in which we get thunderstorm downdrafts.

**Evaporation (liquid to gas).** Evaporation involves a change from a lower to a higher energy state. This is an endothermic process, meaning heat must be removed from the environment.

**Sublimation (solid to gas).** Sublimation is occasionally seen when snow "evaporates". This is a change from a lower to a higher energy state, so it is endothermic and heat must be removed from the environment.

**Melting (solid to liquid).** When snow or ice melts, this is a change from a lower to a higher energy state. Thus it is endothermic and heat must be removed from the environment.

### 1.7.2. Exothermic phase changes

When matter changes from a higher to a lower energy state, it is exothermic. This means that latent heat stored in the water is released into the environment. This release of heat is actually enormously important to forecasting, contributing directly to rain cloud and thunderstorm growth, and we will touch on that throughout this book.

**Condensation (gas to liquid).** The conversion of water vapor to liquid drops is condensation. This change to a lower energy state releases latent heat.

**Freezing or fusion (liquid to solid).** The freezing of liquid water is a change from a higher to a lower energy state. This releases latent heat.

**Figure 1-10. Volcanoes** are a significant source of aerosols, which include mineral dust, ash, and sulfates. This is m,the Cleveland Volcano in Alaska, May 23, 2006, photographed from the International Space Station. *(NASA)*

**Deposition (gas to solid).** The appearance of ice crystals in very cold snow clouds is an example of deposition. This also releases latent heat.

## 1.8. Temperature

### 1.8.1. Environmental temperature

The environmental temperature, $T$, is the actual air temperature observed by a radiosonde or predicted by a numerical model field. It is represented on a skew-T by a vertical line, representing a profile of temperature at many different levels.

### 1.8.2. Parcel temperature

Parcel temperature, $T_p$, differs from environmental temperature, $T$, in that it refers to the temperature in a convective parcel, usually representing buoyant updraft. In most cases the parcel temperature starts with the environmental temperature, and the two depart once lift, subsidence, evaporation, or some other change has been applied to the parcel. It's important in all work with equations and soundings to remember the parcel and environmental temperature are separate.

---

**An early description of latent heat**
The absorption of latent heat is observable in the reduction of a solid to a fluid, as of ice to water, a very great quantity of heat is absorbed before the ice begins to melt, a quantity more than sufficient to bring the temperature above the freezing point, if the heat so absorbed acted sensibly, or could be measured by a thermometer. What then becomes of this large portion of heat, which has incontestibly entered the ice? It has been absorbed by the ice, and lies concealed in it in a latent form.

ANONYMOUS, 1770
*Enquiry Into the General Effects of Heat
(an unauthorized copy of
J. Black's lectures)*

### 1.8.3. Virtual temperature

Dry air is denser than moist air. This is because two of the three atoms making up a water molecule are hydrogen, the single lightest element in the periodic table. By contrast air contains only a trace of hydrogen and is comprised of heavier elements whose masses are comparable that of oxygen. As a result, the molar mass, i.e. mass of 0.602 septillion atoms, of dry air is 0.029 kg while water vapor's molar mass is 0.018 kg.

To take this into account, we use virtual temperature. This is not a directly measured temperature. Virtual temperature indicates the temperature that dry air (devoid of water vapor) must have in order to have the same density as the parcel. *It is important to forecasters because the addition of water vapor makes air behave like it is slightly warmer than it really is.* Dry air is denser than moist air, so the temperature must always increase in order to provide a slightly lower density; therefore $T_v > T$. It can be quickly estimated with $T_v \cong T + r/6$, where $r$ is mixing ratio and $T$ equals Celsius temperature.

The purpose of virtual temperature is to simplify various equations by feeding them an "adjusted" temperature in order to produce air with the same density. If the dewpoint is very low, thus little actual moisture in the atmosphere, the virtual temperature will almost equal the temperature.

But if the dewpoint is very high, indicating significant moisture, the virtual temperature will typically be several K higher than the air temperature, with the difference ranging from about 1 K at subfreezing temperatures to 5 K or more at hot temperatures. A dry atmosphere with its higher mass will need to warm by this amount to have the same density as the parcel.

It is given by $T_v = T/[1-(e/p)(1-\varepsilon)]$ where T is temperature, e is the vapor pressure, and p is the atmospheric pressure, and $\varepsilon$ equals 0.622, the ratio between the gas constant of dry air and water vapor. This shows that when *e* (moisture) is increased, the quantity $1-(0.378e/p)$ decreases, resulting in a higher $T_v$ value.

### 1.8.4. Potential temperature

Potential temperature, $\theta$, or theta, is simply a temperature which is normalized to the 1000 mb level. It is forced along dry adiabats, regardless of moisture. Its temperature at 1000 mb is expressed in Kelvin.

**Figure 1-11. Cumulonimbus towers** contain updrafts undergoing extensive condensation and release of latent heat. *(Author)*

### 1.8.5. Equivalent potential temperature

Equivalent potential temperature, $\theta_e$, or theta-e, indicates the potential temperature a parcel would have if all the latent heat in the parcel was expended and the parcel moved to 1000 mb. This is done on the Skew T diagram by moving the parcel to its LCL and then lifting moist adiabatically to the top of the diagram, then down dry adiabatically to 1000 mb.

### 1.8.6. Subsidence

If a parcel contains condensation, there are two extremes: it can remain entirely within the parcel, or it can be removed completely. This removal process is called sedimentation. The atmosphere is normally in between these extremes, giving what is known as the pseudoadiabatic process. The pseudoadiabatic process is also not reversible, unlike the saturated adiabatic process.

## 1.9. Water

Atmospheric changes would be simple to calculate if the air was devoid of water vapor. Water vapor has three different states at normal atmospheric temperatures, and they each bring different contributions of latent heat, density, and adiabatic lift. These all have a profound influence on the weather.

### 1.9.1. Ideal gas law

The best starting point for the thermodynamics of meteorology is the ideal gas law, first published by Émile Clapeyron in 1834. This is almost as significant to meteorology as $e = mc^2$ is to astrophysicists. The ideal gas law is as follows:

$$pV = nRT$$

where $p$ is the pressure, $V$ is specific volume (e.g. 1 kg of a gas), $n$ is the chemical mass in moles (typically a constant in forecasting), $R$ is the gas constant, $T$ is the temperature,

This can also be rearranged to account for $\rho$, density. Specific density is $V = 1/\rho$ meaning that for a given mass if we increase the density of a substance, the volume $V$ will decrease. This makes sense as a dense substance is by nature

---

**Mesoscale meteorology**

Mesoscale refers to a specific class of sizes in terms of time and space — roughly 5 minutes to 18 hours, and 10 km to 400 km (about 6 to 250 miles). This scale is sometimes called the "subsynoptic" range.

We have a special interest in mesoscale weather because it's much more detailed than the traditional synoptic scale, as seen in the NOAA Daily Weather Map series. It is also a problematic scale, because the basic equations of motion which are used for prediction fail at the mesoscale. There are many more processes that have to be modeled: diabatic heating and cooling, cloud physics, radiation effects, thermal instability, symmetric instability, and much more.

At an even smaller scale than mesoscale we have microscale, which spans the range called (miso and maso by Fujita). This covers 10 km or less, and 10 minutes or less. Many storm-scale processes occur at this range.

Larger than mesoscale is the synoptic scale (called maso by Fujita). This covers anywhere above 400 km or above 18 hours. Most frontal systems, cold fronts, warm fronts, and major highs and lows are in this range.

The ranges above 10,000 km are considered planetary scale.

compact. We can also remove the gas constant. So inserting $1/\rho$ for $V$ and moving it to the other side as the reciprocal $\rho/1$ or just $\rho$, we can simplify this for meteorological use to:

$$p = \rho RT$$

This is an equation occasionally encountered in forecasting, and it shows some important relationships. When pressure is decreased (increased), temperature must also decrease (increase). This is the basis for the observed fall in temperature with height.

### 1.9.2. Vapor pressure

Water vapor pressure partial pressure, $e$, measures the contribution of gaseous water vapor to the total atmospheric pressure in Past mb or hPa. The Earth's atmosphere is a mixture of nitrogen, oxygen, carbon dioxide, water vapor, hydrogen, argon, and many trace molecules like neon, helium, krypton, and xenon. All of these contribute to the total atmospheric pressure, constituting a partial pressure. In meteorology, *vapor pressure refers specifically to the contribution from water vapor.*

Vapor pressure measures the tendency of liquid molecules to evaporate into water vapor at a given temperature. It is also a measure of volatility: chemicals that are volatile have high vapor pressure, like gasoline and ethanol, while substances like oil have low vapor pressure.

Consider that all molecules that are not at absolute zero have internal kinetic energy. Initially at cold temperatures, the kinetic energy is very low, and the substance exists in solid form (i.e. ice). As the temperature of the system increases, molecules are able to break the bonds between neighboring molecules and transition to a liquid. As temperature increases, many of the molecules are able to break away from the liquid into the surrounding air or vacuum in a gaseous state, a process called *flux*.

### 1.9.3. Saturation vapor pressure

Saturation vapor pressure, $e_s$, describes the amount of water vapor pressure that can be added to a parcel before saturation occurs. When this point is reached, the relative humidity equals 100% and theoretically the air can no longer hold any more water vapor. Saturation vapor pressure is a function of temperature.

**Figure 1-12. Fog and haze at the beach** occurs due to the presence of salt which is hygroscopic and provides condensation nuclei. Photo taken near New Haven, Connecticut. *(Author)*

In a parcel, some of the gas constituents convert to liquid, while liquid converts to gas. When the same number of molecules are going into the liquid as the ones going out, the liquid and gas are in a state of equilibrium. The vapor pressure that exists at when the system is in equilibrium is defined as $e_s$, or saturation vapor pressure.

By increasing the temperature further, there will be higher water vapor flux from liquid to gas than the other way around, and the water will lose equilibrium, but as long as the system remains closed it will establish a new equilibrium. This is a new $e_s$, saturation vapor pressure. Thus saturation vapor pressure is proportional to temperature. This relation between temperature and vapor pressure is shown by the Clausius-Clapeyron equation (see margin notes).

## 1.10. The saturation process

The normal state of water in the atmosphere is water vapor, a gas. In forecasting terminology, "moisture" usually refers to water vapor. Also whenever we reference a "moist" parcel, this doesn't mean the parcel is wet or has cloud drops, rather it means the parcel contains some quantity of water vapor, as compared to a dry parcel which is devoid of water vapor.

In a typical atmosphere with a relative humidity of less than 100%, it's a misconception that all the water is in gaseous water vapor form. Even with low relative humidities there are constantly large populations of water vapor molecules condensing into microscopic liquid droplets, while microscopic liquid droplets are constantly evaporating. In dry environments, evaporation dominates, while in saturated conditions, condensation dominates.

In a parcel of air that is not saturated, $e < e_s$ and the relative humidity is less than 100%. Water vapor tends to evaporate. Due to molecular interactions there is a constant supply of extremely small water droplets that develop on condensation nuclei, but they rapidly evaporate. The net state is there is mostly evaporation of water.

When saturation is reached, $e = e_s$ and the relative humidity is exactly 100%. The temperature also equals the dewpoint. The water vapor and the water droplets are now in equilibrium, so this is called the *equilibrium saturation point*. Condensation is likely to form.

---

**A theory of why rain occurs, 1788**

On the one hand, the evaporation of the winter's moisture from the surface of the continent, must tend to saturate with humidity the polar atmosphere, as it acquires an evaporating power from its increasing heat. On the other hand, the progress of the upper current, from the tropic towards the pole, in having its degree of heat diminished by the general cooling cause, will naturally bring the mass to a point of saturation with the aqueous vapor which it had received.

In this state of things, the two opposite currents in the atmosphere, while separate, might pass on without condensing humidity sufficient to produce rain. But the moment that sufficient portions of those saturated streams shall mix, not only cloud, but showers of those saturated streams shall mix, not only cloud, but showers will be produced. Because the sudden formation of a mean degree of heat, in the mixture of two portions in different temperatures, must condense a quantity of vapour sufficient to form rain.

JAMES HUTTON, 1788
*The Theory of Rain*

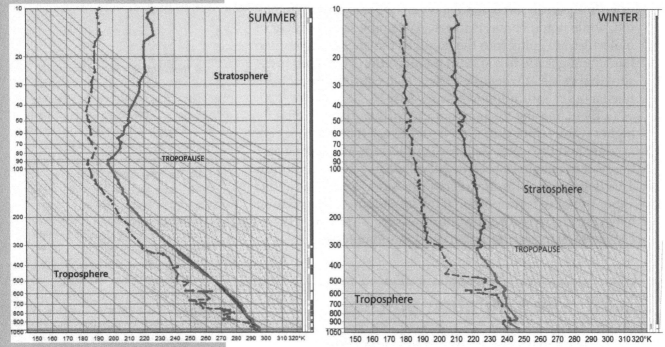

**Figure 1-13. An emagram extending from the surface to the 10 mb level**, about 102,000 ft or 31 km. We see a tropical diagram for Florida in fall (left) and a polar diagram for Wisconsin during the 2014 "polar vortex" event (right). Temperature lines extend directly upward, instead of skewed to the right as with the skew-T diagram, and are labeled in Kelvin  For the warm season case we see a tropopause at 88 mb, or 57,000 ft, while in the cold example it is at 310 mb, or 27,000 ft. This dramatic variation had a major effect on the Concorde supersonic transport during its service between 1976 and 2003. It normally flew at about 52,000 to 57,000 ft, close to the 100 mb level. In the northern winter, warm stratospheric conditions meant crews had to restrict cruise to Mach 1.98 to keep the skin from overheating. In tropical areas, high tropopauses meant a bumpy ride and steering around thunderstorms.

When $e > e_s$ the relative humidity exceeds 100% and the air is considered to be *supersaturated*. Again, there is a constant stream of water vapor condensing into liquid and liquid evaporating back to water vapor. However the parcel is most likely to produce liquid water droplets. The catch is that the water vapor needs surfaces to condense on; the humidity can exceed 100% without any condensation or cloud droplets.

### 1.10.1. Depression

Depression is an expression of the difference between temperature and the property indicated. For example, dew-point depression equals $T-T_d$ and wet bulb depression equals $T-T_w$. Normally the units are Kelvin or Celsius degrees.

### 1.10.2. Saturation methods

Relative humidity can be increased, possibly allowing saturation and condensation to occur, by one of four cooling methods: isobaric, adiabatic, evaporational, and air mass cooling.

**Isobaric cooling**. This is also known as the dewpoint method. With isobaric cooling, the term "isobaric" means constant pressure. So this can occurs without any vertical motion or even any movement. The air temperature is cooled, perhaps by radiation under a clear sky. The decrease in temperature causes $T$ to be reduced, while $T_d$ and $p$ remain constant. In terms of vapor pressure, $e_s$ which is dependent on temperature decreases, while $e$, a function of water vapor content, remains constant. As this process continues, relative humidity increases. When the temperature cools to the dewpoint temperature, saturation occurs. Radiational fog is a common result.

**Adiabatic cooling (uplift).** With adiabatic cooling, the parcel is moved vertically, and it cools at the dry adiabatic or moist adiabatic rate. When the temperature cools to the dewpoint temperature, saturation occurs. It should be noted that mixing ratio, $r$, is conserved, not dewpoint, so the dewpoint actually decreases slightly as the parcel lifts. However once the relative humidity approaches 100%, saturation occurs. Cloud and precipitation development is a common form of this type of cooling.

**Evaporational cooling.** With the evaporational cooling or wet bulb method, water vapor is evaporated into the parcel. Evaporation is a phase change that removes latent heat from the parcel, causing cooling and a change in the temperature. At the same time, the introduction of external moisture into the parcel raises its dewpoint and its mixing ratio. This process continues until the temperature and dewpoint are equivalent. The final values of temperature and dewpoint equal the wet bulb temperature. The production of scud clouds and mist above roads after a heavy rain is associated with evaporational cooling.

**Air mass cooling.** Finally, the air mass method is a process that allows an increase in relative humidity and saturation during the mixing process of two different air masses. For example, take a warm, humid parcel with a temperature of $T=75$ °F and a dewpoint of $T_d=72$ °F, which has a relative humidity of 90% and converts to a vapor pressure of 26.8 mb. It is mixing with an environment consisting of $T=35$ °F and $T_d=33$ °F, which has a relative humidity of 92% and converts to a vapor pressure of 6.4 mb.

When the air mass is half mixed, averaging shows it will have $T=55$ °F and $e=16.6$ mb. However an online vapor pressure calculator shows that the vapor pressure of fully saturated air at 55 °F is 14.8 mb! We have exceeded this by 1.8 mb, and the relative humidity is 114%. So during the mixing process we have produced saturation and cloud material.

**Lavoisier: father of oxygen**

In the 17th and 18th century, scientists speculated on the nature of air and what it was comprised of. In 1703 the German chemist Georg Ernst Stahl built upon work done 36 years earlier by Johann Joachim Becher. Stahl proposed that substances which burn, like wood, coal, and even iron, through rusting, contain a substance called *phlogiston* (Greek "for to set fire"). The phlogiston is supposedly released through the flame into the air.

In 1774, French chemist Antoine Lavoisier (above) made a puzzling discovery. When metals burn or rust, rather than releasing phlogiston, they appeared to gain mass, suggesting that they took phlogiston from the air.

During this same year, British scientist and clergyman Joseph Priestley observed burning substances in a closed container and found that 20% of the air contributed to burning, leaving 80% that is inert. This "fire air" could be synthesized by exposing mercuric oxide to the sun's rays. A mouse was placed in this gas, becoming more lively, while candles burned faster and brighter.

Lavoisier named this gas *oxygene*, from the Greek roots oxy (sharp taste) and gene (producer), both of these originating from a mistaken hypothesis that the gas was present in all acids.

*(Painting: Monsieur de Lavoisier and wife Marie-Anne Pierrette Paulze", oil on canvas, 1788, Jacques-Louis David)*

This is an effect of the Clausius-Clapeyron curve, where the vapor pressure increases exponentially with temperature. You can also use the skew-T to work with mixing ratio instead of vapor pressure to see the same effects.

### 1.10.3. Nucleation

We often discuss water vapor changing to liquid in meteorology, but this overlooks the details of what happens at the microscopic level. *Nucleation* describes any *phase change to a new structure*, such as water vapor to liquid, or liquid to solid, through self-assembly of the molecules. There are two types of nucleation that can occur. Below we refer strictly to the nucleation of water we deal with in meteorology.

**Homogenous nucleation**. The pure process method involves the phase change of only a single component, such as the phase change of a parcel with supercooled droplets of liquid water with no other substances. This is known as spontaneous nucleation, or *homogeneous nucleation*. This is generally only seen in temperatures of below –40°C. It does play a part in the formation of cirrus clouds.

**Heterogenuous nucleation**. There is also the impure process, known as heteregeneous nucleation. Except in the case of cirrus cloud production, this is far the dominant process in tropospheric weather systems. It involves condensation due to the presence of a foreign substance known as an aerosol. The impurity may be a different chemical, such as a sulfate droplet, or it may even be another phase of water, such as a microscopic ice particle entering a parcel filled with supercooled water droplets.

### 1.10.4. Aerosols

Foreign particles which interact with the water are called *aerosols*. Common aerosols include sulfates, nitrates, forest fire combustion particles, fine dust, volcanic dust, sea salt, pollen and even bacteria and viruses. A complete list appears in the margin. In the United States, sulfates make up the most common nuclei, but fine dust is a significant aerosol in the western U.S. Continental areas have about five times as many aerosols as compared to oceanic areas, but in the oceanic regions these nuclei are distributed through a much greater depth.

Sea salt is a common type of nucleus; it is a water-soluble aerosol. Sea salt haze is often observed at the beach,

---

**The effects of oxygen**
The feeling of it to my lungs was not sensibly different from that of common air, but I fancied that my breast felt peculiarly light and easy for some time afterwards. Who can tell but that, in time, this pure air may be a fashionable article of luxury. Hitherto only two mice and myself have had the privilege of breathing it.

JOSEPH PRIESTLEY, 1775
after inhaling isolated oxygen

---

**Sources of condensation nuclei**
**Sulfates** - coal-burning power plants, smelters, and oil refineries. The highest emissions in the US occur south and southeast of the Great Lakes region.
**Carbon** - vehicle exhaust and refueling, cooking, industrial sources, and slash-and-burn agriculture.
**Nitrates** - combustion from vehicles, power plants, industrial sources, and slash-and-burn agriculture.
**Dust** - fine wind-blown crustal material, from the crushing and grinding of sand and rocks. Typically falls out quickly.

giving the characteristic foggy look to a beach during moderate to high humidity conditions.

**Aerosol size**. Aerosols may be as small as 0.02 microns in size, frequently growing to produce larger particles, about 0.2 microns in size, called *cloud condensation nuclei* (CCN). This should not be thought of as comparable to cloud droplets; CCNs are a hundredth the size, close to that of a large virus.

**Wettable and soluble**. Nuclei can also be described in terms of whether they are wettable or soluble. With wettable particles, water spreads out on them as a film. They include many solids, such as dust. Soluble CCNs dissolve chemically into the water droplet. This includes chemicals such as sulfates.

**Distribution**. It is important to be aware of which aerosols are in the atmosphere and where they are abundant. For example over the oceans, there are about 1000 particles per cubic centimeter, but over land this increases to 100,000 in industrial regions. These aerosols are noteworthy because they allow saturation at low relative humidities. This occurs at a level called the deliquescence point, a value that varies according to the composition of the aerosols but is often near 60% relative humidity. This is why various kinds of hazes hang over industrial areas and even over forests: the humidity is just high enough to allow for water vapor to combine with the available aerosols and create a haze.

# 1.11. Moisture properties

We will now list some properties of atmospheric moisture that we work with in meteorology.

### 1.11.1. Mixing ratio

Mixing ratio, $r$, is one of several expressions of actual moisture content in a parcel. It is used frequently in severe weather forecasting for both horizontal maps (alongside dewpoint) and especially in vertical cross sections. Although specific humidity produces a result of kilograms of water per kilogram of air, the meteorological community finds it more convenient to express it in grams of water per kilogram of dry air, g kg$^{-1}$. The value is closely associated with dewpoint

---

**Figure X-XX. Expanded list of aerosol particles.** These are chemicals that are capable of acting as cloud condensation nuclei (CCNs).

**Sulfate** - a molecule of sulphur and oxygen ($SO_4^{2-}$). Sulfates sources are anthropogenic, with 30% being biogenic (e.g. plankton) and 6% are volcanic.

**Nitrate** - a molecule of nitrogen-oxygen ($NO_3^{1-}$). These are mostly anthropogenic; about 28% are biogenic.

**Ammonium** - nitrogen-hydrogen ($NH_4^+$) molecule. Most are anthropogenic; 22% are biogenic.

**Chloride** - a molecule of chlorine ($Cl^-$). It originates from emissions.

**Dust** - windblown inorganics, consisting of minerals and crustal material resulting from erosion. Most is a fine dust, differentiated from the coarse dust particles and sand picked up in duststorms. Some dust is produced by industrial activity. A fraction are caused by meteorites.

**Smoke and soot** - organic combustion particles, particularly that from forest fires, as well as slash & burn agriculture, and fossil fuels. Smoke is organic, unlike inorganic ash. Soot is simply carbon.

**Ash** - inorganic combustion particles that result when heat and oxygen remove all the water and organic matter. It is differentiated from organic smoke and soot. Some ash comes from volcanoes.

**Plant debris** - windblown organic biopolymers, including cellulose. A few percent of global organic aerosols include this category.

**Pollen, spores** - windblown biological aerosols.

**Sea salt** - these are windblown inorganic particles from the ocean.

**Bacteria and viruses** - windblown biological particles. They are used in snowmaking technology (e.g. Snomax).

**SOA** - secondary organic aerosols. These are a large contributor to the "blue haze" phenomenon. SOAs are mostly biogenic but about 4% are anthropogenic, produced by engines and petrochemicals.

temperature, but when a parcel rises or sinks adiabatically, mixing ratio is conserved, not dewpoint.

Some common values of mixing ratio at mean sea level are 3.8 g/kg during dry weather when the dewpoint is 0°C (32°F), and 15 g/kg during a severe weather event when it equals 20°C (68°F). It increases exponentially with increasing dewpoint temperature. If the dewpoint rises another 10 F° to 27°C (80°F), the mixing ratio rises to 23 g/kg.

### 1.11.2. Saturation mixing ratio

The saturation mixing ratio, $r_s$, expresses a property of an air parcel: the maximum amount of water vapor it can "hold". When water vapor is added to a parcel, its mixing ratio increases, and when mixing ratio equals saturation mixing ratio, the relative humidity will be 100%.

Saturation mixing ratio is dictated by the parcel temperature. Some common values at mean sea level are 3.8 g/kg when the temperature is 0°C, and 15 g/kg when it equals 20°C.

### 1.11.3. Dewpoint

In simple terms, the capacity of the air to "hold" water vapor is exponentially proportional to the temperature. Therefore cold air can support only a small amount of water vapor. If dry air is cooled with pressure held constant, there is a certain temperature at which the relative humidity will increase to 100%. This temperature is known as the dewpoint temperature, or simply dewpoint.

If the air is cooled through adiabatic lift, the dewpoint temperature will be slightly cooler than it would be if it were cooled at the original level. Therefore dewpoint is not a conserved property, and forecasters should use mixing ratio or specific humidity to quantify the true amount of moisture in a parcel.

### 1.11.4. Wet bulb temperature

Wet bulb temperature equals the temperature of a parcel of air where the maximum possible evaporational cooling is extracted. This evaporation process removes latent heat of vaporization from the parcel.

Wet bulb temperature, $T_w$, follows the relationship $T \geq T_w \geq T_d$, meaning wet bulb temperature always falls between the temperature and dewpoint. Typically it is slightly closer to the dewpoint than it is to the temperature.

This evaporative cooling can be measured using a wet-bulb thermometer. This is a regular thermometer which has a small sleeve of muslin fitted over the bulb. It is slung in circles by the observer or air is drawn by a fan over it until the thermometer shows no further cooling. This takes about a minute. The displayed temperature is the wet bulb temperature, and from that it can be converted to dewpoint.

Skin can be cooled to wet bulb temperature stepping out of a bath or a pool. Wet bulb temperature is also important to chemical and power plants as it dictates the efficiency of the cooling systems.

### 1.11.5. Equivalent potential temperature

Equivalent potential temperature, $\theta_e$, often referred to by the short term theta-e, represents a parcel in which all possible latent heat is extracted. This is simulated by lifting the parcel to the top of the atmospheric column, rising normally up the moist adiabat from the LCL, then forcing the parcel downward dry adiabatically to 1000 mb. This simulates the process of extracting all of the available latent heat in the parcel. There is no purpose for forcing the air to 1000 mb except to obtain a standardized reference level at a common level. Theta-e is usually given in degrees Celsius.

This expression is very sensitive to water vapor content, and it rises significantly with increasing dewpoint, and to a much lesser extent it is also sensitive to increasing temperature. It is also sensitive to height change; it increases if height rises while temperature and dewpoint are conserved.

Theta-e is often used in severe storm forecasting, and it gives an indication of the buoyant energy in the parcel, though wet bulb potential temperature can provide the same sort of information.

### 1.11.6. Relative humidity

Relative humidity, RH, is an expression of water vapor content versus water vapor capacity in a parcel. It equals the ratio $e/e_s$, $\rho/\rho_s$, $p/p_s$, or (approximately) $r/r_s$, all of them expressed as a percentage.

Contrary to popular perception, a relative humidity value does not define the air's water content. The expression is widely used, however, since evaporation rate of liquids are dictated by relative humidity, which makes it of interest to forecast users like farmers and firefighters. Relative humidity together with the temperature provides an accurate measure of sultriness.

---

**Calculating moisture in 1819**

Of the instruments given to the world under the name of hygrometers, there is not one that shows correctly either the absolute or relative humidity of the atmosphere. The greater part of them depend on the mechanical texture of animal or vegetable substances, and are subject, therefore, to such derangements, as must render their indications altogether unsatisfactory.

The differential thermometer, applied to the purposes of hygrometry, is not liable to the objection now stated; but as its indications are modified by temperature, it does not directly convey any satisfactory information respecting the hygrometric state of the air. An accurate hygrometer, therefore, was still a desideratum in meteorology, till the publication of the article HYGROMETRY, in the Edinburgh Encyclopaedia, where it has been supplied, not be the invention of a new instrument, but from contemperaneous observations of the thermometer and hygrometer, or simply of two thermometers, one of them having its bulb covered with moistened silk or paper [wet bulb].

In that profoud and ingenious treatise, the author has given a formula from which may be deduced, in any given state of a wet and dry thermometer, the following interesting results:
1. The point of deposition, or that temperature at which the air, if cooled down, would begin to deposit moisture [dewpoint].
2. The absolute humidity of the atmosphere, or the quantity of moisture contained in a given portion of air...
3. The relative humidity of the air, complete dryness being denoted by 0, and complete saturation by 100. This result expresses the quantity of moisture which the atmosphere contains, in hudredths of what would produce complete saturation, at the given temperature.

ROBERT GORDON, 1819
*Edinburgh Philosophical Journal*

### 1.11.7. Specific humidity

Specific humidity, $q$, equals $m_v/m$, where $m_v$ equals the mass of the water vapor in the parcel, and $m$ equals the total mass. This total mass is a ratio, comprised of all constituents, the dry air, $m_d$, and water vapor, $m_v$. Although specific humidity produces a result of kilograms of water per kilogram of air, the meteorological community finds it more convenient to express it in grams of water vapor per kilogram of air, g kg$^{-1}$. It is also equal to r/(1+r), where r is mixing ratio, but specific humidity is equivalent to mixing ratio, $q \approx r$.

### 1.11.8. Absolute humidity

Absolute humidity, $\rho_v$, is rarely used in operational meteorology. However it is noteworthy as it expresses the actual amount of water vapor in a parcel, similar to mixing ratio. Units are grams of water vapor per cubic meter, g m$^{-3}$. Since this is a mass per unit volume, it provides density.

It is calculated as $\rho_v = e/R_v \cdot T$ which is similar to the ideal gas equation; $R_v$ equals the gas constant for water vapor. This yields a density, which is the density of water vapor in the parcel.

### 1.11.9. Precipitable water

Precipitable water, $W$, considers the sum of all the water vapor in the troposphere in a column above a given point. So mathematically $\rho_v(z)dz$ is integrated as from $z=0$ to $z$. It gives the mass of water vapor per unit area, kg m$^{-2}$. In forecasting use, this is converted to inches or mm, yielding a value that would fall into rain gauges if all of the water precipitated out.

Precipitable water gives an indication of the mean mixing ratio or specific humidity through the entire column, making it better than the traditional method of checking only a few layers or trying to compare soundings. The technique was actually developed around 1910 by astronomers, for assessing the impact of water vapor on their measurements, but found favor with meteorologists in later decades. Showalter's precipitable water template (1954) and Peterson's precipitable water nomogram (1961) allowed forecasters to compute these quantities on soundings.

Satellite verification studies (Wagner et al 2011) have found the highest annual precipitable water values on Earth are in northern New Guinea and Indonesia and adjacent ocean waters, and in western Brazil. In the northern Americas, the highest precipitable water is found along the northern coast of South America, Central America, and coastal Bay of Campeche.  Precipitable water declines with altitude in nearly all areas, so the driest areas are in Antarctica, Greenland, the Himalayas, and the northern Andes, followed by the Asian-American regions of the Arctic Ocean.

## 1.12. Troposphere and stratosphere

When working with skew-T diagrams, forecasters often deal with features caused by the presence of the stratosphere on the sounding.  It's helpful to know what these features are and where they come from.

### 1.12.1. Troposphere
The majority of weather occurs in the troposphere, the lowest layer in the atmosphere and based on the earth's surface.  Over the United States it averages about 12 km (8 miles) in depth.  The vast majority of weather affecting the surface occurs in the troposphere.  There is extensive vertical movement of moisture and air masses in the troposphere, in part due to the strong lapse rate.

### 1.12.1. Stratosphere
Above the troposphere is the stratosphere, a layer of air characterized by relatively consistent temperatures or even warming with height.  The interface between the troposphere and stratosphere is called the *tropopause*.  The relatively warm conditions in the stratosphere are caused by the presence of molecular oxygen, which together with sunlight produces ozone, releasing heat as part of the chemical process.

Stratospheric layers have traditionally been disregarded in forecasting, as its extreme stability means that vertical motions are suppressed.  For years it was believed that there was no exchange of mass between the stratosphere and troposphere.  However the nuclear testing of the 1950s and 1960s and the sudden appearance of radioactive particles at random locations around the world, known as global fallout, revealed that stratospheric particles were indeed dispersed into the troposphere during tropopause fold events.

Tropopause folds occur when strong vertical motions along the jet stream cause the stratosphere on the poleward side of the jet to sink and advect equatorward underneath the jet stream.  These folds gradually detach and the particles

---

**From 1968**
Precipitable water is defined as the depth of liquid water that would be obtained if the total amount of water vapor above a unit area of the earth's surface were condensed into a layer on that surface. This concept has found major applications in quantitative precipitation forecasting, determination of the moisture flow over an area, and especially in radiation balance studies.

STANTON E. TULLER, 1968
World Distribution of Mean Monthly and Annual Precipitable Water

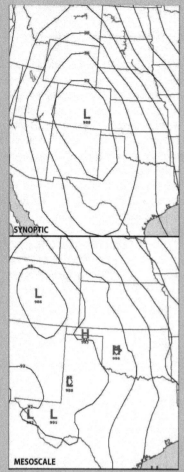

**Figure 1-14. Different analysis scales** of the same weather system, showing deep synoptic-scale low-pressure system associated with the April 10, 1979 Wichita Falls tornado. The top frame uses an analysis filter sensitive to synoptic scale (1000 km) waves. The bottom frame uses a filter that is tuned to mesoscale (100 km) waves. The top frame shows less noise, while the bottom frame reveals troughs that are not apparent on the top chart. (4/10/79 21Z)

mix with the upper troposphere, releasing stratospheric air and any particles that had been carried downward with the fold. Any particles are then quickly carred to the surface through precipitation scavenging.

The tropopause is closely related to the equilibrium level, described elsewhere in this book, which represents storm tops (not the overshooting top). They are not co-located, however. Above the tropopause, there is a significant reduction in cooling or even warming with height, which means that rising thunderstorm updrafts quickly find theirselves cooler than the surrounding air and negatively buoyant. However upper-tropospheric inversions and weak lapse rates can also exist in the troposphere, resulting in storm tops that are lower than the tropopause.

## 1.13. Scales

Finally, a concept we deal with in mesoscale meteorology is *scale*, which describes the dimensions of weather systems in terms of length and time. These two are strongly correlated; for example a tornado has a scale of meters and minutes, while long waves are measured in thousands of kilometers and weeks.

At least 26 known schemes for meteorological scale were developed between 1959 and 1985. The earliest schemes followed a simple principle: *synoptic-scale* covered all meteorological processes that appeared on a conventional weather chart, *microscale* processes contained vertical and horizontal motions with similar orders of magnitude, and *mesoscale* was reserved for all motions in between.

Since those early days, the increasing demands of mesoscale meteorology have required a more careful treatment of the definition. One way to describe mesoscale meteorology is in terms of the equations that are used. Synoptic-scale meteorology can often omit consideration of processes like cloud meteorology, while microscale meteorology can omit effects like Coriolis force. Mesoscale meteorology is unique in that it must consider both sets of equations. This makes it a more technically complex area of meteorology.

In general, mesoscale is best described as having approximate **lengths of 10-100 km (6 to 60 miles), times of 1 hour to 1 day, and velocities of 1 to 10 m s-1 (2 to 20 kt).**

Some journals also refer to the Orlanski 1975 scale (Table 1-2). This provides a division within the mesoscale ranges, with meso-alpha serving as a sort of "sub-synoptic" range and meso-beta providing the classic definition of mesoscale. Meso-gamma is often described as *"storm scale"*.

Credit must also be given to the scale developed in Fujita 1981 which proposed the vowel-based prefixes maso-meso-miso-moso-muso. Although it was not widely adopted, the meso- and miso- boundary of 4 km (2.5 mi) provided a useful threshold value between mesocyclones and tornadic rotation, and the 40 m to 4 km range (130 ft to 2.5 mi) perfectly described the unique scales of microbursts, which Fujita also discovered a few years earlier.

**The many ranges of synoptic scale**
Out of 25 schemes developed between 1959 and 1985 (Linacre 1992), the lower limit of proposed synoptic-scale ranges showed a median value of 500 km (310 miles), with 50% of the values ranging between 200 and 1000 km (125 and 620 miles).

**Table 1-2. Orlanski scale.** Provided here are the scales of motion as described in Orlanski 1975. These are still widely referred to in meteorology resources.

| Scale | Space | Time |
|---|---|---|
| macro-alpha | ≥10,000 | 20d+ |
| macro-beta | 2000-10000 | 2-30d |
| meso-alpha | 200-2000 | 1-5d |
| meso-beta | 20-200 | 2-24h |
| meso-gamma | 2-20 | 0.3-3 h |
| micro-alpha | 200-2000 m | 5-60 m |
| micro-beta | 20-200 m | 0.8-20 m |
| micro-gamma | 2-20 m | 0.1-1 m |

# REVIEW QUESTIONS

1. Describe the difference between sea level pressure (QFF) and altimeter setting (QNH).

2. What type of pressure can an aneroid barometer be set to where it shows the correct pressure no matter where it is transported?

3. What is 0 °C in Kelvin? Write out the unit also.

4. Light is a higher electromagnetic frequency than infrared radiation. Does it have a larger or smaller wavelength?

5. Which type of heat is represented when a hot solid comes into contact with a cold gas, and which way does heat flow?

6. Describe the types of infrared radiation passing through the atmosphere during the day and during the night.

7. What type of heat transfer is represented by molecules of air in contact with hot ground, and what type is associated with cumulonimbus clouds?

8. What type of force drives convection?

9. How effective of a conductor of heat is air?

10. Which type of matter has a fixed volume but whose shape can change?

11. List the three types of phase changes that are exothermic, describing the phase change involved for each.

12. List the three types of phase changes that absorb latent heat, describing the phase change involved for each.

13. Name the two key molecules associated with water vapor, and with dry air, in order of decreasing abundance.

14. Which quantity is indicated by theta, theta e, and theta w, and which level are they referenced to?

15. Describe how equivalent potential temperature and wet bulb temperature are different.

16. The virtual temperature of a moist air mass that has an air temperature of 302 K and a virtual temperature of 306 K.

17. Given the Ideal Gas Law $p = \rho RT$, describe how pressure changes when temperature rises.

18. How does relative humidity change when saturation mixing ratio increases?

19. How does dewpoint temperature vary with a change in mixing ratio?

20. Name at least two expressions that describe the actual moisture content in a parcel, without including contributions from dry air.

# 2 Observing Systems

Although scientists have maintained continuous weather records since the 17th and 18th centuries and developed many early models of surface weather systems, a complete understanding of the atmosphere did not exist until upper-level observations were factored into the weather picture. Our understanding of modern meteorology came only from the exploration of the vertical structures of weather systems deeply embedded in the atmosphere, not just from the limited samples at the base of the layer.

This type of exploration of the atmosphere is actually a relatively recent achievement. It required the development of technologies such as kite trains, balloons, and later on, electromagnetism. It was only in the 18th and 19th century that the first steps were made in reaching higher altitudes and measuring the conditions there with scientific instruments. Only in 1957, thanks to the International Geophysical Year, was there consistency and rapid distribution of radiosonde data worldwide. It was an achievement that even had the full cooperation of the Soviet Union and China, during the height of the Cold War.

All of this made numerical modeling possible. The success of global surface, upper-air, and forecast weather programs in the late 20th century, provided one of the most underrated and impressive achievements of modern humanity, arguably shadowed only by the space program, antibiotics, and the rise of computational technology.

**The 1749 kite ascent**
Upon launching these kites according to the method which had been projected and affording them abundance of proper line, the uppermost one ascended to an amazing height, disappearing at times among the white summer clouds, whilst all the rest, in a series, formed with it in the air below such a lofty scale, and that too affected by such regular and conspiring motions, as at once changed a boyish pastime into a spectacle which greatly interested every beholder.

*Universal Cyclopedia and Atlas, Vol. 6*
*1902*

## 2.1. History

Kites were designed as early as the 5th century BC in China for sending military intelligence and orders across long distances, and in ancent times by the Polynesian people for fishing. It wasn't until 1749, however, that the first weather instruments were sent aloft by Glasgow astronomy professor Alexander Wilson and a student, Thomas Melville. This used a train of six kites to reach altitudes of 3000 feet, attempting to find whether conditions were warmer or colder in cumulus clouds.

Kites were used extensively to sample Canadian air masses between 1822 and 1827 as part of arctic explorer Edward Parry's experiments. These were not meteorological, however, but collected data on temperature and refraction

**Updrafts**
The motions of the kite, whenever a forming cloud came nearly over it, proved an up-moving column of air under it. I speak of cumulus clouds in the form of sugar loaves with flat bases.

JAMES P. ESPY, 1834

**Figure 2-1. US Weather Bureau box kites** were large, expensive to operate. They provided no data until two miles of line were reeled in, returning the recorders to the ground. However it established the first major set of United States upper air data. *(NOAA)*

---

**Stratosphere discovery**
A third distinct stratum has been explored to an altitude of 25,000 meters above sea level. The striking peculiarity of this stratum is that in it the temperature increases from its base upward as far as it has been sounded ... Leading meteorologists differ as to the explanation of this warm stratum.

WILLIAM BLAIR
Exploration of the Upper Air, 1909
Proceedings of the American Philosophical Society, 48, 191

---

just above the earth's surface. This helped polar navigators who needed to use the sextant to sight stars at low elevations.

In 1837, Professor James P. Espy, an early American meteorologist, put forth what is perhaps the first convective index: $z = 100 \times (T-T_d)$, where $T$ and $T_d$ are the Fahrenheit temperature and dewpoint, respectively, and $z$ equals the predicted height of cumulus clouds in yards. In verifying his equation, Espy was able to fly a kite into the base of a cloud 3600 ft above the ground. Espy's work was the first to quantify adiabatic cooling and link it to the formation of clouds and precipitation (Kutzbach 1979). His experiments also showed that air currents existed which moved his kites toward individual cumulus clouds and upward into them.

Numerous kite experiments followed in the middle and late 19th century. The instruments reached heights of several miles with the aid of "booster kites" attached at regular intervals along the line, made of strong piano wire.

The first kite network, providing a vast base of upper-air data, was established in 1898 by the newly-formed US Weather Bureau, starting with 17 kite stations. They were flown from elaborately designed facilities, with three different box kite designs flown for different wind conditions. The kites were tethered with piano wire and attached to a round house on a turntable which faced downwind.

A significant problem with the kites, however, is they could not be flown in weak wind conditions or in bad weather, and were very labor-intensive, requiring up to five hours for a kite flight. Later during the kite program, captive balloons were used in light wind conditions to bring instruments up to measurement altitudes. Another problem with kites is they offered a maximum altitude of about 10,000 feet, or 20,000 feet on very rare occasion, so sampling of the mid-troposphere and the jet stream was not possible. Pilot balloons offered wind information up to a much greater height and with less cost.

Starting with the modernization of the upper air network, the kites were phased out in the United States around 1930, largely due to the increasing risks of aircraft colliding with the steel wire. The weather bureau began contracting with airlines to obtain temperature and humidity data using aerometorograph recording devices that could be bolted to a wing and returned to a local weather bureau office after landing.

## 2.2. Balloons

Hydrogen balloons were known to have been launched and studied as far back as 1783 in France. During the golden age of ballooning that was to follow, small expendable unmanned balloons, or "pilot balloons", were often released before the launch of large manned balloons in order to get information about the winds aloft. Another advent came in 1901 with the invention of sealed rubber balloons, which were an improvement over fixed volume balloons in that they expanded, allowing a constant rate of ascent that made height easy to calculate.

The first balloon-borne instruments were sent aloft in 1892. Most of the launches in the 1890s carried thermographs and barographs. This allowed the discovery of the tropopause in 1902, far above the reach of kites and tethered balloons. However all these devices depended on the payload being found by a citizen and mailed back to the researchers.

World War I provided renewed emphasis on the need for accurate upper-air weather information. Pilot balloons were not reliable at night because they required a lantern and no precipitation or obscuring cloud layers. Kites could not be launched in weak wind conditions, and tethered balloons could not be used in strong winds.

Colonel William Blair at the U.S. Signal Corps laboratory pioneered the first radiosonde in 1924 in Dayton, Ohio, demonstrating radio tracking and the transmission of temperature information to the surface. Pierre Idrac and Robert Bureau in France followed on this work in 1927-1929, and it was Bureau in 1931 who suggested the term "radiosonde", with the U.S. Weather Bureau adopting this word in 1938.

**United States network**

Starting in 1958, the United States radiosonde was designed by the the National Weather Service Scientific Services Division (SSD), and built by VIZ Manufacturing Company. This "A" model received an upgraded hygristor in 1980 and was replaced by a VIZ "B" model at most United States locations in 1988.

In 1988 Space Data Corporation (SDC) radiosondes were assigned to 15 U.S. sites. All of these changes added various levels of bias, error, and inconsistency (Schwartz and Doswell 1991), including high humidity bias in the SDC radiosondes due to the hygristor duct paint being hygroscopic. The SDC radiosondes were used until around 1995 at SEP, DRT, MAF, AMA, ELP, ABQ, TUS, INW, EDW, VBG, GJT, ELY, SLC, LND, BOI, and BIS, and in Alaska at BET, MCG, FAI, and PASY. By the mid-1990s, the US network consisted mostly of VIZ "B" and Vaisala RS80 radiosondes.

In 1997-1998 the network was outfitted with a mixture of Vaisala RS80 and new Sippican B2 radiosondes. VIZ was acquired by Sippican in 1997, and Sippican in turn was acquired by Lockheed Martin in 2004.

The Radiosonde Replacement System in 2006-2008 replaced all radiosonde stations with Sippican MkIIA GPS-equipped radiosondes.

In 2013 the MkII was replaced with the LMS6/LMG6 at nearly all sites. Effectively this changed the carbon to capacitance hygristors.

Some Vaisala RS92-NGP systems are still in use.

15Jun1990
Dayton erroneous sounding

# OBSERVING SYSTEMS

**Radiosonde ascent**

The NWS prescribes an ascent rate of 5.2 m s$^{-1}$ ±12% (about 1020 feet per minute) for radiosonde balloons. Ascent rates that are too slow result in poor aspiration of the sensors, delay the data unneccessarily, and increase the horizontal drift away from the observation site. Ascent rates that are too high indicate an overpressurized balloon, which increases the risk of a balloon bursting midway through the ascent, which contributes to a loss of upper air data. It also uses up the station's hydrogen supply.

dual sites
Medford, San Diego, Bismarck, Del Rio, Washington, Key West, Ketchikan, St Paul, Barrow, Hilo.

**NWS testing**

When the National Weather Service tests radiosondes, it flies the LMS AMT (Advanced Multi-Thermistor) radiosonde, which contains one white thermistor, three silver, and one black, eliminating the effects of infrared radiation. They are then compared using an equation that determines the true temperature.

Meanwhile Pavel Molchanov flew the first Soviet radiosonde in January 1930 from what is now St. Petersburg, Russia.

Starting in 1936, a radiosonde network was deployed in the United States and gradually expanded through the 1940s and early 1950s. However there is great inconsistency in the data quality and observing practices during this time, and this limits the ability to use data from this period for severe weather studies.

The modern radiosonde network slowly came together during the 1940s, and went through a period of rapid grown in 1947. The modern standardized 0000 and 1200 UTC observations began in July 1957 with the beginning of the International Geophysical Year (IGY). Radiosonde coverage then gradually increased and peaked in the 1980s. Since that time, there has been a slow decline due to budget cuts, competition with other agencies for government funds, and the increasing, sometimes questionable use of numerical model data as a proxy for observed data.

## 2.3. Pilot balloon

The pilot balloon, or PIBAL, is the simplest upper air instrument, and it remains in use in certain countries for obtaining cloud heights and upper-level wind information. The pilot balloon is a hydrogen or helium balloon that is released and tracked visually. Modern pilot balloons have a constant ascent rate, as long as they aren't accelerated by convection. This makes it possible to estimate the altitude at any given instant. They are still used today in some countries where cost prohibits the use of radiosondes.

Starting in the 1920s, pilot balloons were used extensively, replacing the kite. The simplest use of them was to determine ceiling height of low clouds. A weather observer simply released a balloon and watched its ascent with a stopwatch. The time elapsed provided the ceiling height.

Designated pilot balloon stations used a more complex arrangement to obtain wind information aloft. Observers used a theodolite, a calibrated telescope with azimuth and altitude scales which was kept trained on a balloon by an observer while the data was logged. By combining this with elapsed time giving height, it was possible to compute the coordinates of the balloon, and from that, the wind direction and speed was known.

In the United States pilot balloons were inflated with hydrogen to a few feet in diameter and then released. These balloons could be tracked by a good observer to altitudes of 32,000 to 48,000 feet during the day, as the balloons grew in size in the thin air. They could be launched at night with a lantern or electric light, but seldom could be tracked higher than 10,000 feet at night (U. S. Weather Bureau 1931). They provided wind data, but did not provide temperature or moisture information.

## 2.4. Modern radiosondes

The vast majority of atmospheric soundings since the mid-20th century have been taken using radiosonde observations. The 21st century has brought a substantial increase in the amount of additional data, from profilers, radar velocity data, satellite remote sensing, dropsondes, and forecast model output. However radiosondes remain the backbone of the world's upper air data.

The modern radiosonde is about the size of a small milk carton and weighs about 8 ounces (240 g). These instruments are launched twice a day at about 800 locations worldwide. They are released by a human observer and carried aloft on a hydrogen balloon made of latex or neoprene. The sensor package is suspended about 20 to 60 m beneath the balloon in order to avoid contaminating the sensors with solar heat and moisture captured by the balloon. Radiosondes rise at a rate of 5 m s$^{-1}$, crossing through the troposphere over the period of about 45 minutes. They penetrate deep within the stratosphere to an altitude of about 30 to 35 km, covering a ground track of a few km to 250 km before it bursts. The radiosonde then falls to earth, slowed by a parachute.

We will now go into the construction of radiosondes. Forecasters should be familiar with these as they define the instrument's behavior, response to measured values, and their susceptibility to unusual weather conditions.

### 2.4.1. Flight train

The flight train refers to the grouping of instruments and accessories that make up the radiosonde, as they are all assembled on a single nylon separation line, much like the cars on a freight train. The balloon rides at the top of the

**Accuracy requirements**
The information here summarizes the accuracy requirements for upper-air measurements used in synoptic meteorology, as published by the World Meteorological Organization. These values are for levels between sea-level and 100 mb:
**Pressure**: 1 mb; 2 mb near 100 mb
**Temperature**: 0.5 K (0.5 C°)
**Relative humidity**: 5%
**Wind direction**: 5°; for winds of 30 kt or higher, 2.5°
**Wind speed**: 1 m s$^{-1}$ (1.9 kt)
**Geopotential height**: 1% at the surface decreasing to 0.5% at the tropopause

**What happens to radiosondes**
Radiosondes are normally packaged with instructions that allow finders to send the unit back to the weather agency free of charge for refurbishment and re-use. In the United States, only about 20% of the radiosondes are ever recovered. Most of these discoveries are by farmers and ranchers. The rest are lost in trees, woodlands, and ocean waters, or are mistaken for trash.

All parts of the radiosonde contain non-biodegradable material, and efforts to make them environmentally friendly have not been successful so far (WMO 2017). The polystyrene material used in the housing does not biodegrade, and resists photolysis and oxidation. It is believed the plastic degrades after about 50 years but the plastic itself could take thousands to millions of years to decompose.

**Upper air status in 2017**
Radiosonde coverage has been in slow decline worldwide since the 1980s due to budget cuts, the expense of GPS radiosondes, and reliance on numerical model fields. A snapshot in September 2017 showed the following:
**North America** - 70 stations in the conterminous US (44570 $mi^{-2}$ per station) made up of 68 NWS stations, one DoD site (VBG), and one DOE site (LMT); 13 in Alaska; 2 in Hawaii. Weak coverage in Canada with 28 sites (32% the density of the US). Large data void in Mexico with only 2 sites. In the Caribbean 11 sites on island nations. No Central American sites.
**Europe** (excl. Russia) - Good coverage with 62 stations (110% the density of the US). No coverage in Spain, Greece, the Baltic States, and Belarus.
**Africa** - Poor coverage with a total of 10 sites for a continent covering 12 million square miles; four were in West Africa. Virtually no coverage in East Africa.
**Middle East** - Iran 4; Saudi Arabia 3; UAE 1; Israel 1.
**Asia** - Good coverage in China with 82 sites (95% the density of the US) with large voids in and around Tibet. Numerous stations in Russia. Good coverage elsewhere: Japan 18; Indonesia 16; Malaysia 9; India 6; Philippines 6; South Korea 5; Mongolia 4; Vietnam 3; Bangladesh 2; Taiwan 1. Coverage is nonexistent in Burma, Thailand, Laos, and Cambodia.
**Australia** - 19 sites covering the continent, yielding 25% the density of the US.
**South America** - 43 sites (27% the density of the US). Brazil's network is the most developed. Nearly all countries operate stations. No contributions from Paraguay, Bolivia, and Venezuela. No coverage in far southern Argentina and Chile. Many stations only report at 1200 UTC.
**Antarctica** - 8 stations (6% the density of the US), with a heavy coastal bias.

---

flight train and the radiosonde at the tail end. Other accessories, however, may be provided in the middle.

The train has to be designed with the idea of minimizing stress on the balloon neck, as well as avoiding collisions, entanglements, and contamination. There is also a maximum length permitted on the flight train, 120 ft in the United States, as beyond that a pendulum effect will occur that can cause signal dropout. A typical length is about 85 ft in the United States. A minimum length is also necessary to prevent the balloon from affecting the radiosonde package, and to move the radiosonde out of the wake of the balloon, where it can be difficult to dissipate absorbed radiation.

**Balloon**. The balloon rides at the top of the flight train, providing the upward lifting force. It is made of latex or synthetic rubber, and is filled with hydrogen or helium. The goal is to achieve an ascent rate of 5 m $s^{-1}$ (300 m $min^{-1}$).

When initially filled, the radiosonde balloon is about 1.85 meters (6 feet) in diameter. It grows as it ascends, reaching about 10 meters (30 feet) wide before bursting at an altitude of about 30-35 km. After this happens, the parachute becomes the top component of the flight train if one has been added.

**Parachute**. A parachute is connected underneath the balloon. Its purpose is to manage the descent rate after the balloon bursts, slowing the rate of descent of the flight train to less than 10 m $s^{-1}$ (22 mph). This minimizes the risk of damage or injury once the radiosonde reaches earth.

The parachute is colored bright orange so the radiosonde can be spotted by pilots during ascent and descent. The color also improves the chance the radiosonde will be discovered and sent in for refurbishment by citizens. Parachutes are omitted at remote stations where the risk of a falling radiosonde striking people or property is negligible.

**Light**. Some balloons may carry a glow stick or an LED light. This is used to allow pilots to spot the balloon, including during the parachute descent. It is also used for some ground stations to help aim the antenna during the first few monites of flight.

**Shock unit**. A shock unit is an elastic clip, spring, or line that may be added to the cord to minimize instrument er-

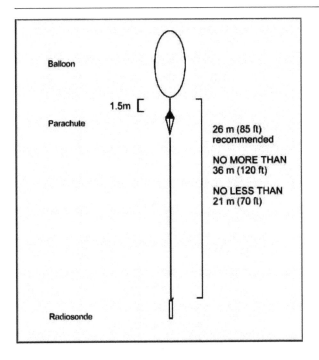

**Figure 2-2. Schematic of a radiosonde flight train**, provided in the U.S. Federal Meteorological Handbook No. 3. *(US Office of the Federal Coordinator of Meteorology)*

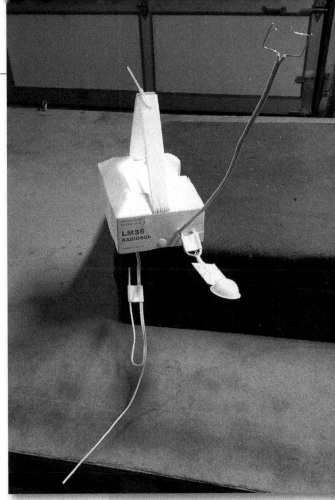

**Figure 2-3. Lockheed LMS6 radiosonde** launched from Denver in November 2017. We see the temperature sensor at top. The white upside down cup in the center protects the humidity sensor from precipitation. The antenna is at the lower left. *(Courtesy Alycia Gilliland)*

rors. The shock unit prevents sharp jerks from whipping the radiosonde around in strong winds..

**Flight bar.** Dual or triple radiosondes, which are used only for research or testing, may be added to the cord. This splits the flight train into two or three units, allowing additional radiosondes to be suspended beneath. This keeps them from swinging into each other or interfering with one another.

**Radiosonde.** Beneath the parachute is the radiosonde. All of the instruments within the radiosonde are contained in an enclosure made of polystyrene. The radiosonde is prepared for flight and a water-activated battery is used. The water-activated battery works for about two hours, and besides the plastics it is the source of chemicals of greatest concern

### 2.4.2. Data link

While it is ascending, the radiosonde transmits a one-way radio signal with measurements. This is known as telemetry. To perform this, an electronic transducer in the radiosonde converts all the sensor readings electronic signals.

42  OBSERVING SYSTEMS

**Hair hygrometers**

Hygrometers which use human hairs are one of the oldest modern meteorological instruments. Hairs contain keratin, whose molecular hydrogen bonds can be disrupted by the presence of water vapor, and this causes the hair to lengthen. The design was first published in 1783 by Genevan earth scientist Horace-Bénédict de Saussure. This type of hygrometer was used on European radiosondes from the 1930s until the 1970s.

It can be a fun project to build such an instrument since materials are readily available. A design using ordinary household items and a glue gun can be found at: <www.exploratorium.edu/exploring/hair/hair_activity.html>. Other designs can be created using a carton or small box for a frame, rubber bands for tension, and thread to position the hair. Laser pointers with a miniature mirror can be used to project a dot on a scale.

---

The data channels of pressure, temperature, humidity, and so forth are multiplexed into a single signal. Time-division multiplexing is the most common format, where each instrument gets a slice of signal time, and this forms a packet of data transmitted once every 1 to 6 seconds. This multiplexed signal is then added to the radiosonde's carrier signal.

The carrier signal can be on one of two bands internationally allocated for radiosondes: 400-406 MHz (referred to as UHF, or 403 MHz), and the shorter-wavelength higher-frequency 1675-1700 MHz (referred to as SHF, 1675 MHz, or 1680 MHz units). The former UHF system is more common, while the latter one is shared by weather satellites.

The short wavelength of 1680 MHz radiosondes are subject to much more attenuation in the atmosphere than the 403 MHz radiosondes. This means that they're more likely to fail in situations with strong atmospheric winds, which could carry the radisonde to low elevation angles where they are hard to detect.

The telemetry signal is received at the launch location by a ground station. Here it is decoded and assembled by a computer. The data is then formatted into standardized messages for distribution to the weather community. Much of this had to be done by hand before the 1980s and it was a time-consuming process. In developing countries, many ground stations are still not automated due to budget problems, and the data may have to be telephoned to a longline facility for transmission. Transcription errors are frequently introduced. Forecasters using this data have to be especially attentive in monitoring integrity of the sounding, and modify them as appropriate.

### 2.4.3. Pressure sensor

Pressure is the most critical element of the radiosonde since all the other data elements depend on its accurate measurement. The pressure sensor uses an aneroid ("without fluid") cell, which relies on an elastic response to atmospheric pressure.

Older instruments that were common during the mid-20th century used an aneroid metal bellows, which caused a mechanical arm to move across a commutator plate. This position was measured electrically and encoded as a signal by the circuit. They have been made obsolete by miniaturization.

Newer designs use a miniature micromechanical aneroid cell. Pressure is measured with a silicon capacitance

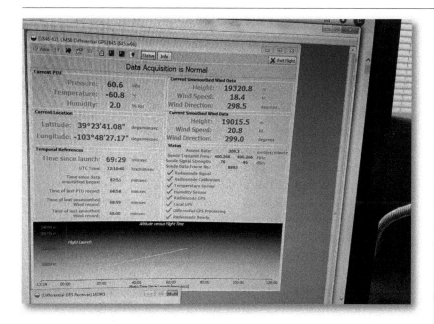

**Figure 2-4. LMG6 ground station display** of a Lockheed LMS6 radiosonde in flight reaching an altitude of 19 km (63,000 ft). Note that a little over an hour has elapsed during this flight, as the balloon rises at 5 meters per second. *(Courtesy Alycia Gilliland)*

sensor, in which plates inside the evacuated aneroid capsule move closer or further apart, changing the cell's ability to store electronic charge on the plates. This is measured and encoded by the circuits. It provides pressure with minimal moving parts, and has excellent hysteresis, responding to pressure changes in less than a second. The Vaisala RS80, RS92, and LMS radiosondes all use this type of capacitive aneroid cell.

All aneroid cells are prone to leaking once they are reach high altitudes, usually above 15 km but sometimes lower, which causes errors with the calculated heights. The radiosonde output shows an abrupt changes in pressure when this happens. However the ground station has the ability to detect and compensate for most of these situations before the observation is transmitted.

This pressure value is converted to height once the sounding is assembled using the hydrostatic equation. This provides the basis for height measurements. Typical height errors are about 10 meters (about 1 mb).

### 2.4.4. Temperature sensor

Temperature is directly measured, and there are three key technologies which can be used in radiosonde temperature detection: capacitance (electrical), resistance (electrical), thermocouples (electrical), and bimetallic strip (mechanical). Regardless of which sensor is used, it is positioned on the outside of the radiosonde so it detects the ambient air.

**Old radiosonde problems**

Starting in 1973 the United States, which was using NWS SSD radiosondes with carbon hygristor sensors, began the process of inserting bogus humidity data whenever humidity for a level was determined to be less than 20%.

This was done because of accuracy problems with the sensors which were identified in 1973 (Quiring 1973, Nordahl 1978) and later found to be due to defects in the humidity calculation algorithms and in the clock-driven radiosonde chart recorder (Wade 1994). A NWS memorandum abolished the practice on 1 October 1993.

Unfortunately the replacement of data with bogus information made statistical reconstruction impossible. As a result, upper-level moisture information for the United States between 1973 and 1993 for dry layers has been lost permanently. Climate change research is the primary area that has been impacted by the problem.

> **Kite experiments in 1822**
> It appears by an experiment that, when the sea is covered with ice in the winter, there is no sensible difference between the temperatures of the atmosphere at the surface of the ice, and at the height of 400 feet above it. This was tried by means of a paper kite with an excellent register thermometer attached to it, the altitude of which was determined by two different observers at the time, at a given distance from each other, and in the same vertical plane as the kite, and from which the perpendicular height of the kite above the level of the ice was computed. This experiment was tried under favourable circumstances, at the temperature of -24° Fahrenheit. The kite was sent up and caught in coming down without the thermometer being in the least disturbed, the indices of which did not show the slightest alteration, although carefully compared before and after the experiments and the kite remained at the same height for a considerable time.
>
> GEORGE FISHER, 1825
> *Appendix to Parry's Journal of a Second Voyage*

**Figure X-XX. Thermometer coating types.** Listed here are typical emissivity and solar absorptivity values.

| Coating | Emissivity ($\varepsilon$) | Absorp ($\lambda$) |
|---|---|---|
| alumin. | small/0.22 | mod/0.31 |
| white | large/0.86 | small/0.12 |
| black | large/0.86 | large/0.94 |

**Resistance**. The electrical thermal resistor, or thermistor, is the simplest and cheapest method for measuring temperature. It can be provided on the radiosonde as a thermistor or a resistance wire. It relies on the unique properties of a semiconductor, usually a metallic substance. A change in the temperature increases or decreases the amount of electrical voltage or current passing through the circuit. Voltage can be thought of as water pressure (psi) and current as water volume (gph), and either can be affected.

Typically radiosondes use an NTC (negative temperature coefficient) thermistor, consisting of a chip of metal oxides. It has high resistance at cold temperatures, and decreases as temperature rises. The response is nonlinear, which can introduce accuracy problems, and the resistance becomes exponential at high temperatures.

The sibling of NTC thermistors is the PTC (positive temperature coefficient) variant. It has low resistance at cold temperatures, increasing resistance with height. These are not common in radiosonde design. They are made of ceramics or barium compounds.

One design problem with all thermistors is self-heating, as passing current through a resistor converts it to heat; the same principle used by electric stovetops and radiators. However the low voltage and high resistance used in radiosonde thermistor design tends to make this warmth negligible in practice.

Thermistors came into common use in radiosondes starting in the 1950s. They are the most common method of measuring temperature. The Viz/Sippican/Lockheed B2, Mark II, and LMS6 sensors, the latter currently making up the backbone of the U.S. network, are examples of radiosondes using thermistors. Older designs used iron fillings in a housing of baked clay.

**Capacitance**. The thermal capacitor contains a capacitor that responds to temperature changes. This tends to have low hysteresis, responding rapidly to changes. Older designs used a sensor sealed in glass, but a twin-platinum wire design is now common, separated by an insulator. This creates a field that passes current depending on the temperature, and if it is not passed through, it accumulates in the capacitor as charge. The radiosonde has a circuit that monitors the characteristics of this charge and converts it to a temperature reading.

Capacitance temperature sensors first came into widespread use on radiosondes in 1981 with the Vaisala RS80. They have become the backbone of the Vaisala product line, and are known by the trade name "Thermocap". Capacitance sensors were also introduced to German Graw radiosondes in the 1990s.

**Thermocouple**. A thermocouple joins two different metals together. The environment heats the metals, and electromotive force is produced as thermally energized electrons move through the thermocouple. The current is negligible, but the voltage is measurable and is converted by the radiosonde to temperature. The weak voltages may be a source of error, but the response is linear, which simplifies the design.

The main advantage of thermocouples is that they are cheap to build and very rugged. Radiosondes that use the thermocouple are limited to the Swiss MeteoLabor series, which was introduced around 2005. These radiosondes are in use currently.

**Bimetallic strip**. A bimetallic strip sensor is a mechanical device which binds together two pieces of metal with different coefficients of expansion. With a certain amount of heating, one of the metal strips expands much more than the other, causing it to warp or deform by a certain amount. This deformation can be measured with a transducer arm, which converts it to an electrical signal. From that, the temperature can be determined.

Bimetallic sensors were very common on early radiosondes, but they were gradually abandoned, and dropped entirely when Vaisala shifted to the RS80 in 1981. Due to the cheap and reliable design of modern capacitance and resistance sensors, which have no moving parts, bimetallic sensors are unlikely to return to use in radiosondes.

**Latent heat errors**. The most significant problem for radiosondes is when the instrument emerges from cloud layers into a clear air above, particularly when rising from wet cumuliform clouds into very dry air. Radiosondes may show cold biases, producing cold spikes and layers with absolute instability. This is caused by evaporative cooling of wet sen-

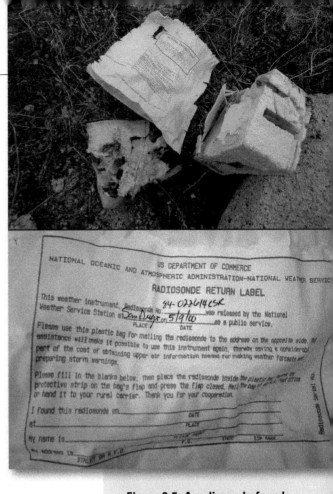

**Figure 2-5. A radiosonde found in the mountains of southern California**, found by a hiker 11 years after launch. The radiosonde drifted about 50 miles east-southeast of the launch point at Miramar NAS and returned to the ground by parachute. Return rates in the United States are in the 10 to 20% range; the rest are discarded as trash or are never found. *(Courtesy Daren R. Sefcik)*

**Figure 2-6. Vaisala RS80 radiosonde**, the most widely-known radiosonde of the 1980s and early 1990s. The capped element sticking out under the flap is the capacitive hygrometer and temperature sensor. The gold packet is a water-activated battery that is connected to the 9V connector near the top. A barometric sensor is within the instrument, not seen. At the very top is an antenna and flight train connector. The batteries are packaged with a 1-foot strip of punched tape used to calibrate the ground station. *(Author photo)*

sors, causing them to cool, possibly all the way to their wet bulb temperatures.

Old designs solved this by housing the temperature element in a duct to protect it from rain, though this increases the risk of stagnation of the air sample. Current Lockheed LMS6 radisondes use a shield, protecting the radisonde from drops above. Vaisala radiosondes rely on a water repellant coating.

In the 1970s rare warm spikes were observed on U.S. radiosondes, about 10 mb deep and ranging up to 2°C in amplitude. They were determined to be caused by supercooled water in cumuliform clouds that froze to the thermistor, where latent heat of fusion was transferred (Pietrowicz and Schiermeier 1978).

**Radiation errors.** Another problem is contamination by shortwave and longwave radiation. During the day the sensor can be warmed by shortwave radiation from direct illumination by the sun, the ground, clouds, and even the balloon itself. The rod shape is also vulnerable to a warm bias from sunshine at low angles. Cold biases can also come from radiosondes within clouds, where the normal radiative exposures are absent.

At night, stray radiation is not a problem. However the radiation of longwave excessively from the radiosonde into the surrounding environment can produce significant cold biases. This is especially a problem with high-emissivity thermometers, which are common in meteorology.

To minimize these problems, radiosondes are designed to have low emissivity, high reflectivity, and low absorption characteristics. Low emissivity means that radiation is emitted at a slow rate, and biases due to infrared contamination are minimized. However this increases the risk of warm biases in unusual conditions.

Until the 1990s, thermometers often had a white coating, and it was found that they often showed a cold bias at night and a warm bias during the day, with errors up to 1.8 °C. The adoption of aluminum (alumized) coatings, which have low emissivity, made them immune to radiation contamination, but makes it difficult for them to dissipate their absorbed heat and can result in daytime warm bias.

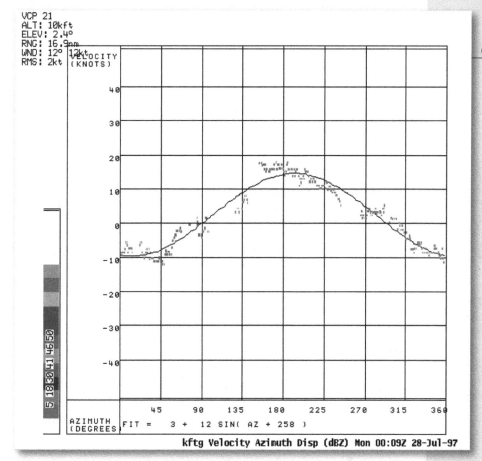

**Figure 2-7. VAD velocity graph.** This shows a sample of radar data for a specific range from the radar, representing a specific height. Note that the wind information (dots) forms a sine wave. By fitting a sine wave on the wind data, we obtain the wind data for that level, represented by the peak of the fitted sine wave. In this case, the winds are blowing from 010° toward 190° (out of the north) at a velocity of 15 kt.

Older white coatings actually had problems with high emissivity, causing significant heat exchange errors.

Errors are also minimized by algorithms developed by each radiosonde manufacturer which are coded into the ground station. Many algorithms consider solar angle, cloud situation, and so forth in order to model the radiation correctly. All of this reduces errors to within 1 K on average.

**Hysteresis errors.** Temperature sensors are also susceptible to hysteresis (lag), especially if the mass of the sensor is large. Older radiosondes from the middle of the 20th century had signicant hysteresis, while newer units respond almost instantly. The different lag characteristics remains a challenge for researchers studying old data and comparing it with new data. However for forecasters using today's radiosonde data, this is not an issue unless the radiosonde goes through a sharp thermal gradient.

**Instrumentation and processing errors.** Occasionally errors can be produced if the radiosonde is damaged, or introduced

**Dropsondes**
The counterpart to the radiosonde is the dropsonde, which has a special role in hurricane forecasting. A "Hurricane Hunter" aircraft will drops radiosonde sensor package from flight level on a special parachute canopy. The dropsonde slowly falls to earth at a speed of about 50 kt, gradually slowing to about 20 kt near the surface. As it falls, it continuously transmits data back to the aircraft building a sounding from the top down.

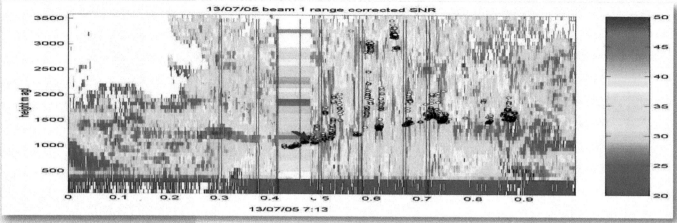

**Figure 2-8. Research profiler in Europe**, showing a cross section with 3500 m (11490 ft) at the top. This shows the base of a summertime stratocumulus layer rising with time. *(Courtesy UK Met Office)*

by the ground station processing equipment. Common errors include insufficient modelling of radiative processes. The Vaisala RS80, which was used extensively in the 1980s and 1990s, used an aluminum sensor that underwent an excessive software correction, producing a slight warm bias of about 0.25 K, with errors increasing in the stratosphere.

### 2.4.5. Humidity sensor

Most of the humidity sensors on radiosondes use a humidity sensor that changes electrical resistance or capacitance. This has historically been the most unreliable measurement, and has created much trouble for climatological researchers trying to assimilate old radiosonde data into long-term models. Moisture plots from unfamiliar parts of the world should be analyzed with caution as the humidity sensors may be deficient or have different characteristics compared to US and European networks.

**Goldbeater's skin.** This is an older system still used in Russian designs, and in previous decades it was used in China and eastern European radiosondes. Goldbeater's skin changes in length according to the water vapor in the atmosphere. It relies on the hygroscopic properties of ox intestine, which is soaked in potassium hydroxide or salt, washed, stretched, beaten or pressed flat, and treated to prevent decomposition.

This material has immense tensile strength. Because of this, it was used as a protective layer to beat gold into microscopically thin sheets, and was used in Zeppelin airships as gas bags. When used in radiosondes, it has good humidity

OBSERVING SYSTEMS    49

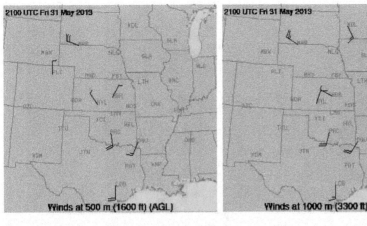
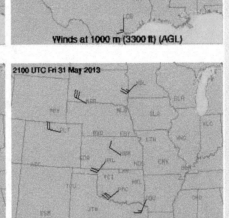
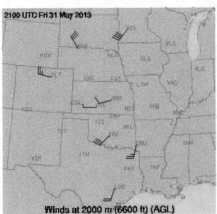
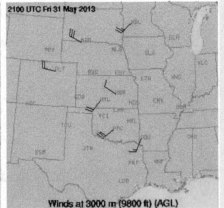

**Figure 2-9. National Profiler Network** plot of winds aloft, shortly before it was shut down, on May 31, 2013 a couple of hours before the El Reno EF4 tornado event. The winds shown here can be used to construct an updated plot of winds aloft.

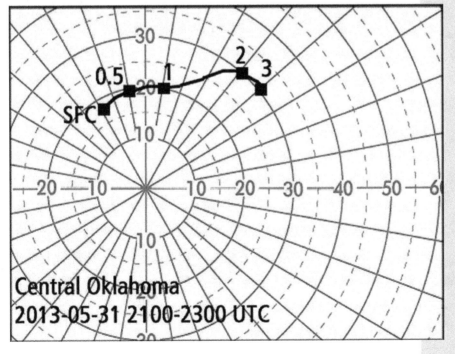

**Figure 2-10. Hodograph of the data shown above** with the surface data for the Oklahoma City area. This does not show the classic curved hodograph with the hook shape, indicating that mesoscale processes likely shaped the hodograph west of Oklahoma City. A mesolow in the Binger-Anadarko area would have slowed down the low-level flow or backed it to more easterly, which would have added the necessary curving. METAR observations at Yukon's Clarence E. Page airport did show backing to 130°, which would have elongated the surface-to-0.5 km segment and significantly increased storm-relative helicity.

50 OBSERVING SYSTEMS

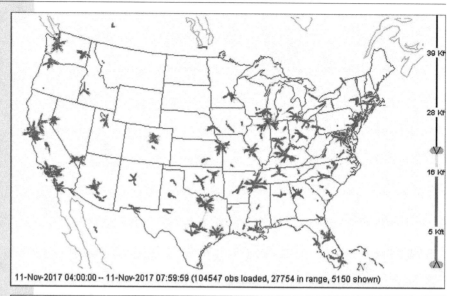

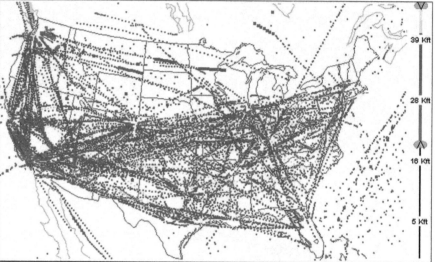

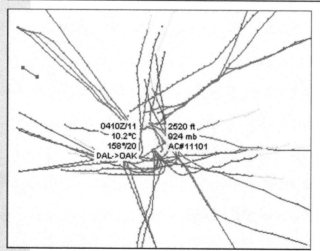

**Figure 2-11. Four hours of ACARS data** plots for November 11, 2017, showing measurements from the surface to 20,000 ft (top) and 20,000 to 45,000 ft (center). This shows the very strong bias for lower and mid tropospheric measurements being situated close to airport terminals, while upper tropospheric observations have a more even distribution across the country. Colors correspond to altitude. Also shown (bottom) is a sample of one plot from the arrival and departure pattern at DFW International Airport, showing an aicraft reporting winds of 158° at 20 kt at 2520 ft MSL. Such data, if accessible, can be useful in fine-tuning hodograph data.

response at mid-range humidity values, but has slow response at high and low humidity levels, often overestimating dry and moist conditions. It also has problems with hysteresis (lag) and errors below -20°C.

**Lithium chloride element**. One of the oldest types of humidity sensors is the lithium chloride element. As with many other types of sensors, it detects the change in electrical resistance. Lithium chloride reacts with moisture in the atmosphere to change the amount of current it allows to pass through.

Lithium chloride sensors were introduced in 1943 in the United States radiosonde network, and were replaced by the carbon hygristor around 1965. Lithium chloride also enjoyed use as part of U.S. ground station humidity sensors from 1960 to 1985 as the HO-61 hygrothermometer.

**Carbon hygristor**. The carbon hygristor uses a film of carbon black (elementally pure carbon powder) mixed with a hygroscopic binder, and mounted on a glass or plastic strip. This forms an electric resistor. To measure humidity, the radiosonde determines its resistance to current. The carbon hygristor was introduced in the United States in 1965, improved in 1980, and used until the 1990s. It was used in the Viz/Sippican B2 and Mk II that was widely used in the US in the 2000s and early 2010s.

Carbon hygristors are prone to developing defects once they become saturated in a high-humidity environment, and the instrument is often in error for the remainder of the flight. Even exposure above 60% tends to show a moist bias. They also show hysteresis when emerging from the tops of clouds, that is, they respond slowly to sudden temperature changes, and they show significant errors in low relative humidity or below -40°C. The Sippican MkII that was in use in the US also showed a dry bias in dry conditions.

**Capacitive hygrometer**. A relatively new technique is the use of capacitive microsensors. Electrodes are placed on either side of a porous polymer which soaks up water vapor, and this is mounted to a glass substrate. This method is now used in the American radiosonde network on the Lockheed LMS-6 radiosondes. Capacitive hygrometers are believed to be more accurate than the carbon hygristor sensors that were previously used by American radiosondes. The capacitive hygrometer has also found its way into the U.S. ASOS official

---

**An aside on MDCRS**

Though I rarely editorialize in my books, I encourage readers with the right connections to work within their organizations to press the airlines to remove restrictions on MDCRS data. Considering that aircraft movements are now widely available on sites like FlightAware and FlightRadar24, the policy of making this data private, which was established at the outset of the ACARS program in the mid-1990s, is antiquated and runs against the core concept of sharing scientific data.

Data sharing allows advances to be made through interdisciplinary use of the data, facilitates long-term preservation, and allow the data to be easily used as training tools for researchers and new meteorologists. Again, the changes need to come from the airlines and their working groups.

Author

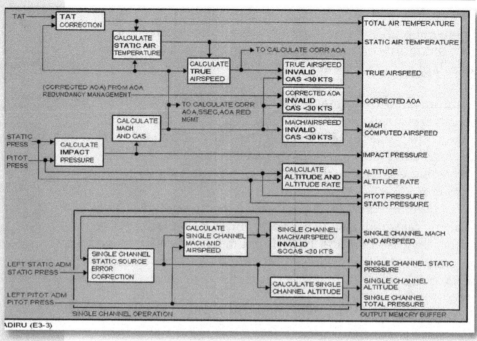

Figure 2-12. **Air Data Inertial Reference Unit** for a Boeing 777 airliner, showing the complex flow of data used to compute outside air temperature, pressure, altitude, and other data from sensors along the nose of the aircraft (left). *(Continental Airlines)*

**Radiosonde history**

Although sensor-equipped kites were widely used in operational meteorology in the early 20th century, the very first modern radiosonde as we know it dates to 1929, when M. Idrac and M.R. Bureau tested the Thermoradio that transmitted temperature data to a ground station. The first network radiosondes date to 1936 when Professor Vilho Vaisala of Finland began mass production of radiosonde units and shipped the first units to MIT in Massachusetts. It was only during the 1950s that a global upper-air network became established.

ground network as the DTS1 unit, replacing the HO-1088 chilled mirror sensor in the mid-2000s.

Capacitor sensors were introduced by Vaisala as the "Humicap" (humidity capacitor) and came into widespread use on the Vaisala RS80 and RS90 in the 1980s. The RS80 entered the United States observation network between 1994 and 1998.

The thin-film capacitor is a variation used in American designs like the LMS6, which forms the current US radiosonde network. It has good performance, comparable to the Vaisala RS92 twin thin film capacitor.

**Chilled mirror.** The most accurate sensor available is the chilled mirror hygrometer, where water vapor condenses on an artificially chilled surface. A photodetector detects condensation of fine water droplets. This directly measures the dewpoint temperature. The technology was used extensively from 1985 to 2004 in official U.S. stations as the HO-83 and in ASOS surface sites as the HO-1088, but due to its maintenance costs it was replaced with a capacitive hygrometer around 2004.

Due to its expense, chilled mirror hygrometers are not used in normal radiosondes, but it serves as the moisture technology for the Meteolabor "Snow White" radiosonde, which is used as a reference standard to test the humidity sensors of other radiosonde equipment.

Even with the chilled mirror method, the problem of the instrument rising out of the cloud into dry air may cause measurement problems.

### 2.4.6. Wind measurement

The measurement of wind is perhaps the most difficult problem in upper-air observation. Radiosondes cannot directly measure wind, since they are embedded within the wind field and observe calm ambient conditions. An anemometer carried on board would simply report calm winds or turbulent eddies. Therefore wind measurement is based on tracking relative to a ground reference point.

Early systems used a radiotheodolite, which uses radio direction finding (RDF) to lock onto the telemetry signal with a small steerable dish antenna. The pressure combined with the azimuth and elevation makes it possible to determine the radiosonde's x/y/z position. The changes in this position over time can be be used to calculate displacement, thus, we obtain the wind field. These "radar radiosondes" are still in use in Russia.

One problem with RDF techniques is in high wind conditions, such as those observed around the polar front jet and the low-level jet, which can cause the radiosonde to ascend at low angles relative to the horizon. This increases the possibility of radio reception problems such as multipathing.

It is also possible to use navigational radio systems such as GPS. Some older radiosondes used the Omega system (decommissioned 1997) and LORAN-C system (decommissioned 2015), which were ground-based positioning systems. These eliminate the need for a steerable dish. However by shifting position measurement onto the radiosondes, this increases their cost, typically by about 50-75%.

Some use both RDF and GPS, and are known as dual mode systems. They allow the ascent to be constructed from RDF reception if there is a problem with the GPS signal. The United States uses a dual-mode system.

The base pressure, temperature, and humidity data makes the observation a radiosonde observation, but the addition of wind data makes the observation a rawinsonde observation (radio wind sonde). However "radiosonde" tends to be used for both types of observations.

---

**WWII radiosondes**

Study of captured radiosonde equipment now allows comparison with that used by our country. The German and Japanese use techniques and measuring elements similar to ours but differing in the type of elements and in the method of transmitting the information. The Germans use two general types of radiosondes. One employs wet and dry-bulb mercury-in-glass thermometers for temperature and relative humidity measurements and a mercury-filled glass manometer for pressure measurement. Two transmitters and antennas and two radio frequencies are required. The Germans also use chronometric radiosondes employing bimetallic elements for temperature and hair hygrometers for humidity measurements. Japanese radiosondes are very similar to those of the Germans.

German and Japanese radiosondes are well-designed and constructed and are smaller in size and lighter than weight than our own equipment. However our radiosondes are the only ones designed for mass production. In addition, our radiosondes give a greater number and consequently more accurate readings than those of the enemy.

QST MAGAZINE, February 1945
Amateur Radio Relay League

### 2.4.7. Data output

The radiosonde ascent data with every sample of data is stored in an internal format. The resolution of this data ranges from 1 second for the current Radiosonde Replacement System (RRS) to 6 seconds for the older Micro-ART system. However a version usable by the meteorological community is sent out over weather networks, using the WMO TEMP-35 and PILOT-32 formats. This is then ingested by weather software to plot soundings, by model analysis components, and can be plotted manually by forecasters on paper soundings. A higher resolution format called BUFR (Binary Universal Form for the Representation of meteorological data) is slowly gaining acceptance.

In the United States, the radiosonde is required to be released within a one-hour window starting 30 minutes before the synoptic observation time (NOAA 2010). Typically the balloon is released at the start of this time block and observations are available shortly after the synoptic hour.

It should be kept in mind that the TEMP (TTAA/TTBB/PPBB) format is a condensed version of the high-resolution sounding data. Only wind direction changes of 10° or 10 knots, temperature changes of more than 1° C past the interpolated temperature, or relative humidity changes of more than 15% from the interpolated relative humidity, are considered significant enough to report (NWS 2010). This may allow some subtle temperature, humidity, and wind structures may go undetected. The BUFR format provides a higher-resolution picture of the temperature profile, but as of 2017 this was not being disseminated at most stations.

**Figure 2-13.** Aircraft like this provide an important source of MDCRS data, filling in the gaps between radiosonde sites. The main drawback is the lack of moisture data and the lack of data in the lower half of the tropopause between major airports. Shown here is AA MD-80 N482AA at the gate at DFW Airport, April 6, 2006. *(Author photo)*

## 2.5. Radar velocity data

Doppler radar units like the WSR-88D are a very under-utilized source of wind data. The radar provides base products which include reflectivity, velocity, and spectrum width, along with polarimetric measurements. Many products are derived from them, ranging from storm identification to hail detection.

### 2.5.1. Velocity azimuth sampling

As the radar beam forms a conical shape, it is able to sample wind velocity at many different levels. By taking data at a specific range and elevation from the radar, we can obtain wind data at a specific height.

Unfortunately we can only measure the component of motion toward or from the radar. Assuming the wind field is uniform, if we scan a specific height in a full 360° sweep and graph the velocity data, we will see the wind velocity form a sine wave (Figure XXX), with the peak of the sine wave representing wind direction blowing directly away from the radar, and the points on the zero line representing winds blowing perpendicular to the radar.

By fitting an actual sine wave to the data, we obtain the wind information — winds are blowing from the azimuth represented by the low point of the fitted sine wave, and blowing toward the peak. The amplitude of the sine wave gives us velocity.

**Figure 2-14. Developing Oklahoma thunderstorms** looking from the rear, showing the arrangement of updraft towers along a discrete multicell line. *(Author photo)*

### 2.5.2. Velocity azimuth display

By collecting all of the processed velocity azimuth data, we can build a profile of the wind direction and speed with height. This is often plotted on a velocity wind profile (VWP), where the X-axis represents time with older data to the left or right, and the Y-axis represents height. This allows forecasters to see trends in the wind direction and speed over time.

It is important to remember that this is based on the wind field being uniform. If there are errors where many data points do not match the fitted curve, the measurement at that level is considered to have a high probability of error. By convention the wind barb is plotted with a yellow or red shade.

Also when there are not enough scatterers to reflect the radar energy, the level will be considered to have no data. This is represented by an "ND" on the display.

The velocity wind profile is very useful for mesoscale forecasting as it gives the wind just above the surface. Hodographs can be adjusted with this information. The main problem is that levels are only provided every 1,000 ft (304 meters), so the critical tornado shear levels under 1 km are coarsely sampled.

## 2.6. Wind profiler

Profilers are a kind of ground-based radar specialized for measuring wind data aloft. The antenna points upward at a slightly tilted angle. The United States profiler uses three beams, one of them vertical and the other two at right angles to each other. The radar measures the frequency shift of backscattered radar power. A major advantage of the wind profiler is it can continuously monitor the wind field, whereas radiosonde data is only available every 12 hours on average. As with weather radars, wind profiler performance is closely related to their wavelength. Microwave (1300 MHz) profilers are strongly attenuated and can only reach about 5 km, while 400 MHz UHF profilers can sample the entire troposphere up to about 15 km, and 50 MHz VHF research profilers can reach well into the stratosphere to heights of 30 km.

In the United States, the National Profiler Network, consisting of 404 MHz UHF units, was deployed primarily in the central US and Mississippi Valley region. The system had been shown to improve severe weather watch accuracy and lead time on tornado warnings. The United States NPN network ceased operation by September 1, 2014, concluding that the ACARS data from aircraft provided the same information.

Elsewhere, a network operated by the Earth System Research Laboratory is in use in the western United States, with several units fielded in the Denver and Vermont area. Europe, Japan, Ontario, and Quebec also operate networks of wind profilers.

Profiler information is a supplemental source of data, providing wind information between the 12-hour radiosonde samples. Therefore it can detect quasigeostrophic disturbances and unexpected strengthening of the jet stream. It is also useful for adjusting hodographs and soundings and evaluating changes in the boundary layer.

OBSERVING SYSTEMS   57

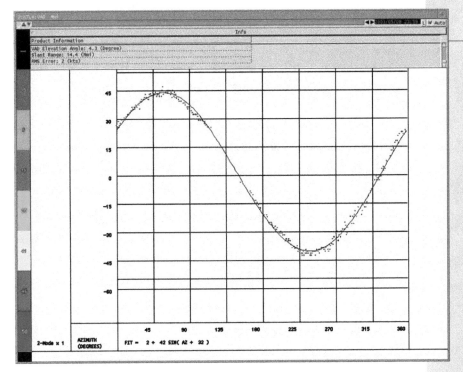

**Figure 2-15a. WSR-88D velocity azimuth display (VAD) plot.** This is almost never available for examination, but this sample gives some idea of what is involved in computing a profile. This shows information for a given tilt. The x-axis is azimuth, and the y-azimuth is the velocity, with returns away from the radar (positive velocity) above zero, and returns toward the radar below it. The dots show individual gates (samples) received in all directions. The sine wave shown is fitted, providing the wind profile for that tilt. If the gates vary from the sine wave, this indicates uncertainty. Many departures from the sine wave will result in a larger RMS error that will be plotted in VWP (below) as shades of yellow and red.

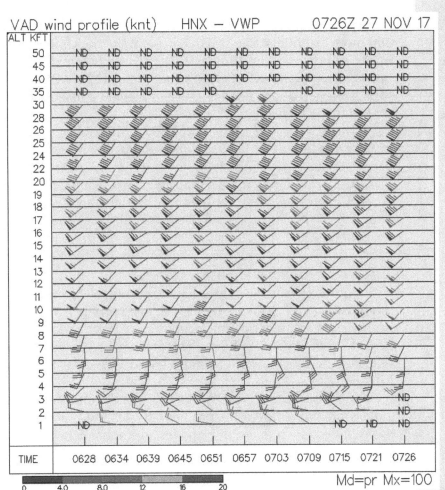

**Figure 2-15b. WSR-88D velocity wind profile (VWP) plot.** This product is widely available on internet sites, including College of DuPage (shown here), an excellent site for weather analysis. They are also available on weather.gov, WeatherTap, These products give an excellent indication of minute-to-minute wind changes aloft. Higher RMS error numbers (shades of green, yellow, and red) in this color scheme indicate a lack of homogeneity in the velocity field, and usually indicates the presence of a front, a mesolow, or a quasigeostrophic disturbance, or some other mesoscale system that is causing the wind field to vary in specific areas. It can be an excellent signal to look more closely at the data. The actual velocity frames are a good starting point.

*(Author photo)*

## 2.7. AMDAR / ACARS / MDCRS

Data reported from aircraft is known as AMDAR, Aircraft Meteorological Data Relay. This is transmitted automatically from mostly passenger flights. Aircraft flight management systems have the capability to sense outside temperature and calculate wind information based on heading, airspeed, and ground speed. There has been work to also include water vapor sensors. This data is sampled every 6 to 300 seconds, depending on the phase of flight, and is then sent via data link to a network of ground stations.

### 2.7.1. Configuration

AMDAR is a worldwide system of aircraft data reports, most of them from European airlines. In the United States, the AMDAR system is operated by ARINC (Aeronautical Radio Incorporated) using a network called ACARS, Aircraft Communication Addressing and Reporting System. Just to add more acronyms, the AMDAR program in the United States is known as Meteorological Data Collection and Reporting Service (MDCRS). Altogether there are about 100,000 MDCRS observations per day, most of them at cruise altitude (25,000 to 41,000 feet).

The first ACARS test was in 1979. AMDAR began appearing worldwide in 1992, and in the United States MDCRS began service in 1997, involving the National Weather Service, ARINC, and several major airlines: American, Delta, Federal Express, United, United Parcel Service, and Northwest, which was absorbed into Delta. Southwest joined the program in 2005.

### 2.7.2. Meteorological data

Although extensive information is available at cruise altitude between 25,000 and 41,000 feet, deep profiles through a large layer are only possible near large airports as aircraft descend and climb from them. This is the main drawback with both AMDAR, which includes worldwide data and U.S. MDCRS data.

Some benefits of AMDAR include the obvious: fine-scale wind and temperature profiles between the 12-hour radiosonde launch times. It provides additional data to be ingested into high-resolution models. It helps forecast wind gusts and temperature inversions. In severe weather situations it helps modify hodographs, giving a better picture of severe weather potential, and can provide information on changes in the capping inversion.

### 2.7.3. Drawbacks

One problem with AMDAR is that the data is not as accurate as that coming from radiosondes, which are extensively tested. A recent study showed a warm bias in AMDAR, pointing out that the errors may already be contaminating model data (Ballish and Kumar 2008). However it found that most observed temperatures errors were limited to 1° C or less. Also the unavailability of moisture data limits its use in skew-T log-p analysis, but can provides some data on changes in inversions and caps.

The other problem is that MDCRS data is restricted to federal government computers. International AMDAR data worldwide is distributed, however, and appears in the upper-air files posted by university servers. Tools like Digital Atmosphere will plot this data where it appears. However ACARS-derived data does not appear in these datasets.

## REVIEW QUESTIONS

1. What was the earliest system for observing temperature aloft?

2. Name the earliest convective weather index, developed by Professor James P. Espy in 1837?

3. How does an observer determine height from a pilot balloon?

4. How often are balloons released as part of the modern radiosonde network? What are the launch times in UTC?

5. What is a flight train?

6. What is the typical maximum altitude for a radiosonde?

---

**AMDAR questions**

**How many data are there?**
Currently, we are getting just about 140,000 wind and temperature observations per day, 100,000 of which are over the continental United States. These data come from more than 4000 aircraft. There are more data during the daytime than at night, but thanks to participation by some parcel-carrying airlines, nighttime coverage is substantial.

**What kind of altitude is reported?**
The altitude-determining physical variable is pressure. This is converted to altitude by using a standard atmosphere. The standard atmosphere is used at all altitudes and under all pressure conditions. Thus, for instance, it is possible to have ACARS altitudes below ground level on days with high atmospheric pressure. (This is in contrast to other aviation reports such as voice PIREPS that use a standard atmosphere to compute altitude above 18,000 ft (MSL), but use the current altimeter setting for lower altitudes.)

**What is the accuracy?**
Benjamin, Schwartz, and Cole (1999, Weather and Forecasting, in review) -- collocation study with ACARS reports. Estimated wind vector accuracy - 1.8 m/s, estimated temperature accuracy - 0.5 deg C. These numbers are much lower than the previous ACARS/RAOB differences, which by definition included RAOB error.

7. How does the radiosonde send measurements to the ground station?

8. What is a disadvantage of the 1680 MHz SHF (super high-frequency) radiosonde?

9. How does the mechanism work that detects pressure in a modern radiosonde?

10. Name the two technologies used in radiosonde thermometer sensors.

11. What is the most significant problem for temperature measurement?

12. Name a technology used nowadays to detect water vapor.

13. What causes the problem with radiosonde data in high winds?

14. What is the old radiosonde data format called, and what is the newer radiosonde data format?

15. How might radar velocity data be used for sampling the pre-storm environment?

16. What is the most significant advantage of wind profiler data?

LMS6 display. *(Alycia Gilliland)*

17. Are wind profilers capable of detecting trends in storm-relative helicity?

18. Are basic wind profilers capable of detecting changes in the temperature layers within a capping inversion?

19. AMDAR is a program for obtaining upper-level data from which system?

20. What is the main problem with both ACARS and AMDAR data?

Vaisala RS80 displays. *(Author photo)*

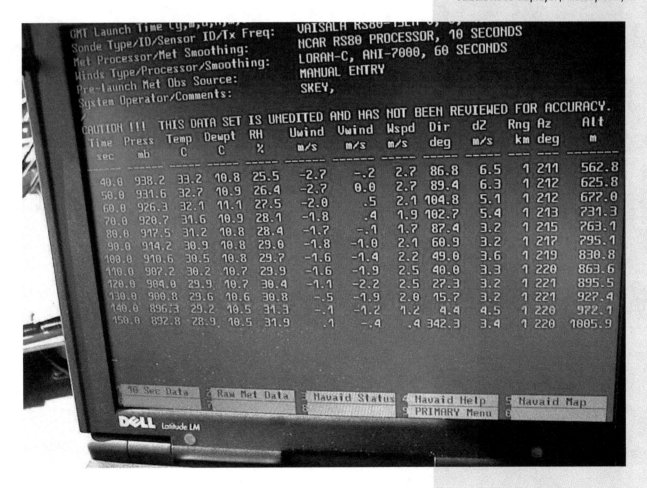

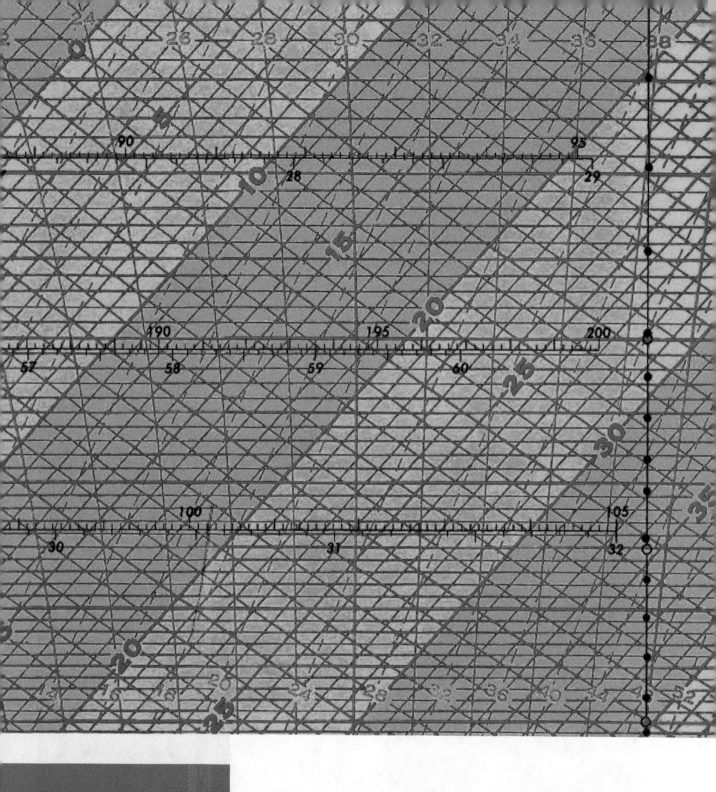

# 3  SKEW-T LOG P DIAGRAM

The 18th century and the early 19th century established the foundations of the thermodynamic diagram. Physicists James Clerk Maxwell, Ludwig Boltzmann, and Rudolf Clausius had mapped out some of the earliest equations relating to heat, pressure, and moisture. As the middle of the 19th century passed and the United States recovered from the Civil War, meteorologists already understood the role of phase changes and entropy in weather, and were developing tools to assist with the analysis and forecast process.

By the 1860s, a telegraph network stretched across the United States. This meant that observations at stations thousands of miles away could be plotted within minutes. This rapid communication of data was the key that made *operational meteorology* possible. During these early years, the U. S. Signal Corps was responsible for charting out the nation's weather, and began keeping track of the movement of weather systems using plots of pressure, wind, temperature, and cloud cover.

By the 1880s, forecasters were increasingly in need of a graphical diagram to represent the complex adiabatic processes taking place in the atmosphere. Text tables were readily available, but these were awkward and time-consuming for forecast use. Finally in 1884, the German physicist Heinrich Hertz demonstrated a graphical method for plotting processes involving convection, phase changes, and subsidence on a chart. This chart was the emagram, the forerunner of all soundings.

Improvements on the emagram followed a few decades later. One of them was the Stüve diagram, in 1927 by German meteorologist Georg Stüve. It resembles the emagram but straightens most of the lines for a clearer presentation, while sacrificing use of geometric-area methods for energy computations. But in 1915 British meteorologist Napier Shaw developed the tephigram, an radical development as it skewed the temperature lines against the slope caused by the temperature fall with height. This made important changes much easier to see.

In 1947 the skew-T log-p diagram was developed, which straightens the pressure lines, this being an important detail to differentiate it from the tephigram. The skew-T was created by Norwegian meteorologist H. Herlofson. It was adopted by the United States and has become widely used. As the tephigram is similar to the skew T log-p diagram, nearly all information about skew-T's in this book also applies to the tephigram.

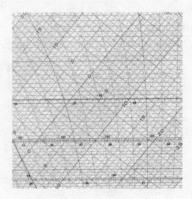

**Summed up from the beginning**
The theoretical meteorologist daily has to discuss considerations to the changes in condition that take place in moist air that is compressed or expanded without the addition of any heat. Hence he desires to attain answers to these questions with the least possible expenditure of time, and he does not care to use any of the complicated formulae of thermodynamics.

DR. HEINRICH HERTZ, 1884
*The Mechanics of the Earth's Atmosphere*

**History of the Skew-T**
The configuration used for the Skew-T log P diagram is credited to Nicolai Herlofson, who proposed a variation of the emagram in a 1947 article published in *Meteorologiske Annaler*.

64   SKEW-T LOG-P DIAGRAM

**Figure 3-1. Heinrich Hertz's emagram from 1884**, the godfather of all modern Skew T log p diagrams and tephigrams. (*The mechanics of the earth's atmosphere*, Trans. by Cleveland Abbe, Smithsonian Institution, 1891)

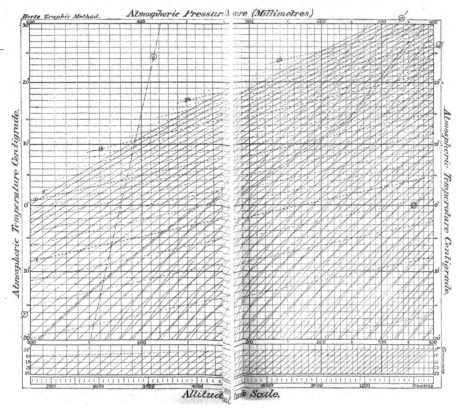

**Figure 3-2. Probably the earliest available weather analysis**, showing weather conditions across the Midwest and eastern United States on January 1, 1871. This appears to show a progressive polar outbreak pattern with an Alberta clipper and cold front stretching from Michigan to Illinois. The invention of the electric telegraph a couple of decades earlier made this possible. *(NOAA)*

## 3.1. The skew-T log-p chart

The skew-T log-p diagram, along with the tephigram, emagram, and the Stüve diagram are all examples of a thermodynamic diagram. They all depict the same relationships, with the same temperature lines, adiabats, and pressure lines, except the coordinates are arranged differently between each type of chart. They provide a visual representation of the first law of thermodynamics, which maintains that energy can be transformed from one form to another, but can be neither created nor destroyed.

When using an unfamiliar chart, such as that found in old journals, all you need to do is determine the coordinate system the chart uses, then do your work on the chart accordingly. All thermodynamic digrams used in meteorology are based on one particular configuration: showing increasing temperature along the abscissa (X-axis) and increasing altitude on the ordinate (Y-axis). The only difference between various types of diagrams is the coordinate arrangement, some of which provide curvature, slope, and other characteristics.

The United States with very few exceptions uses the skew-T log-p diagram, commonly known as the skew-T. Since North America is the target audience for this book, we will focus entirely on the skew-T. Other countries use either the skew-T or the tephigram.

The altitude range is typically 1050 to 100 mb (about −990 to 53100 ft geopotential altitude), and these are widely available on the Internet. Low-altitude diagrams are used by some forecasters, such as the 1050 to 400 mb diagram (−990 to 23600 ft), highly useful for parcel calculations, or the 1050 to 10 mb (−990 to 102,000 ft) which captures weather in the stratosphere. Special diagrams exist which contain refractivity overprints; these are used by radar technicians.

## 3.2. Construction of the skew-T

In this section we explain all the lines found on a skew-T diagram. It's important for new forecasters to learn the configuration and orientation of these lines, because many journal articles and slideshow presentations omit the labeling

---

**Quick sounding identification**

**Skew-T log-p:** Isobars (x) are horizontal and *straight*; isotherms (y) are skewed 45 degrees up and to the right.

**Tephigram:** Isobars (x) are horizontal and *curved*; isotherms (y) are skewed 45 degrees up and to the right.

**Stuve:** Isobars (x) are horizontal and isotherms (y) are vertical. Dry adiabats slope up and to the left, and are *straight*.

**Emagram:** Isobars (x) are horizontal and isotherms (y) are vertical, Dry adiabats slope 45 degrees up and to the left and are *curved*.

---

**Standard intervals**

**Pressure:** 50 or 100 mb for major lines. 10 mb for minor lines.

**Temperature:** 10 °C for major lines, 2 °C for minor lines.

**Dry adiabats:** 10 °C for major lines, 2 °C for minor lines.

**Wet adiabats:** 10 °C for major lines, 2 °C for minor lines.

**Mixing ratio:** Logarithmic: major values include 0.1, 0.2, 0.4, 1, 2, 4, 7, 10, 16, 24, 32, 40, 48, 56, and 68 g/kg.

**Thickness:** If used, 1000-500 mb, 700-500, and 1000-700 mb are typically included. Additional options may include 500-300, 300-200 mb, and others as needed.

of various lines, and reference values may be left out, leaving the forecaster to guess what they are.

Observations of temperature on the diagram are placed on the diagram using for coordinates temperature and pressure $(T, p)$. This is all that's needed to place a piece of data. Observations of moisture are plotted according to their dewpoint and pressure $(T_d, p)$. Either the temperature or the moisture profile can be omitted and this is often the case when a sensor fails, but this usually prevents adiabatic lift from being calculated and prevents the use of many indices.

### 3.2.1. Isobars (pressure)

Pressure lines are drawn horizontally, and are spaced logarithmically to provide a constant change of geopotential height and satisfy the area-energy relationship. On the tephigram the isobars show unmistakable curvature. Typically this line is black or white.

Rising from the bottom to the top of the diagram we always see decreasing pressure and thus increasing height, since the atmospheric pressure diminishes with height. Printed diagrams will show standard geopotential height in feet (parentheses) and meters (brackets) near the isobar labels.

The bottom of a normal Skew-T is 1050 mb (about –990 ft MSL) while the top of the diagram is normally 100 mb (about 53,000 ft MSL). With specialized plotting software like RAOB it is possible to generate soundings with other top and bottom values.

### 3.2.2. Isotherms (temperature)

Temperature lines are vertically-oriented, but in the case of the Skew-T diagram they slope up and to the right, thus the name "Skew-T". Since the natural tendency of the atmosphere is for temperature to decrease with height, this helps the plotted data stand more erect, and allows the forecaster to more easily spot important changes which would otherwise be lost in a slanted line. This line is normally black or white.

Some diagrams will contain a Fahrenheit scale at the bottom of the chart. This identifies the temperature value for the isotherm lines which intersect with the bottom of the chart and the scale.

---

**On tornadic storms, 1942**

The lapse rate in the cold Marine Polar air mass to the westward of the warm, moist Maritime Tropical massi s about the same as the lapse rate in the Marine Tropical mass to the eastward, and has about the same temperature up to some 1.5 km, above which the lapse rate is considerably steeper in the Marine Polar mass [post-dryline air] than in the Marine Tropical mass to the eastward. During the day the Polar mass, which has really become a Superior mass due to subsidence, becomes warmer in its lower levels up to 3,000 or 4,000 ft than the Marine Tropical air mass ahead of it. Above the level where the lapse rate in the Marine Polar mass begins to steepen sharply, the air becomes increasingly colder than the Marine Tropical air ahead of it, usually being much colder in the higher levels. This colder air from the intermediate and upper levels in the Marine Polar mass is usually dry, and flows out over and above the lower portion of the Marine Tropical mass, cutting off the top portion and lifting it very rapidly, due to the steepness of the cold front aloft, and violent thunderstorms and occasional tornadoes develop where vertical convection is strong enough.

J. R. LLOYD, 1942
U. S. Weather Bureau
The Development and
Trajectories of Tornadoes

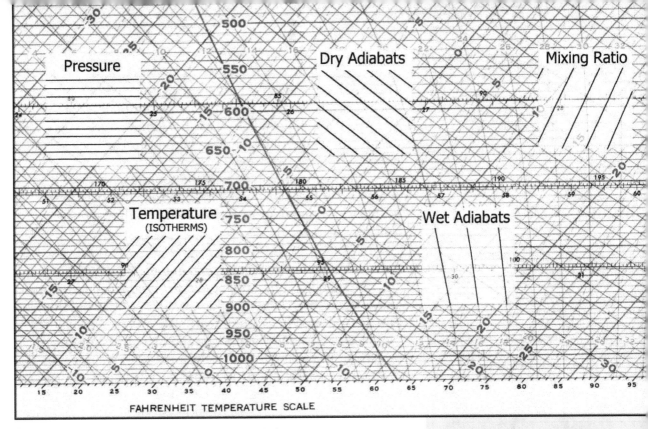

Figure 3-3. Skew-T log-p components, showing the dry adiabats (isentropes), the moist adiabats (wet adiabats, or pseudoadiabats), the pressure lines (isobars), temperature lines (isotherms), and mixing ratio lines. The mixing ratio lines can always be differentiated from the moist adiabats because the latter lines are curved, and the other are straight on the skew-T.

### 3.2.3. Dry adiabats (potential temperature)

Dry adiabats, also known as isentropes, or subsaturation adiabats, slope up and to the left. This line is normally a "dry" color like brown or red.

These are lines of equal potential temperature ($\theta$, theta). These lines depict the cooling and warming a parcel undergoes when it ascends or descends adiabatically (the DALR, or dry adiabatic lapse rate). By adiabatically, this means without any phase changes and without mass or heat added to or removed from the parcel to the outside environment. The units are degrees Celsius where they intersect the 1000 mb reference level; some special diagrams will use Kelvin. So to use the diagrams in conjunction with isentropic analysis, forecasters must be able to readily convert between Celsius and Kelvin (K=°C+273.16; °C=K−273.16).

Parcels that are on the same dry adiabat share the same potential temperature. That is, if two values on the same adiabat and move vertically to a common pressure level without any exchange of heat or moisture, they will meet and their temperature will be identical.

### 3.2.4. Moist adiabats (wet bulb pot. temperature)

Moist adiabats, also known as pseudoadiabats, wet adiabats, or saturated adiabats, are curved vertical lines.

### Radiosondes

The NWS uses two radiosondes: the Lockheed Martin LMS6 and the Vaisala RS-92NGP. The LMS6 product line was originally a VIZ / Sippican product.

# 68  SKEW-T LOG-P DIAGRAM

They are almost always drawn with a green color. These are lines of equal wet bulb potential temperature ($\theta_w$, theta-w). The values for moist adiabats are the temperature where the line intersects the 1000 mb reference level. The line is colored green.

The moist adiabat depicts the cooling and warming a parcel undergoes when it ascends or descends with latent heat being added to or removed as the result of a phase change.

### 3.2.5. Mixing ratio lines

The mixing ratio lines are vertical, sloped slightly up and to the right. They are ruler-straight, normally plotted in dashed or dotted lines, and are colored green. A moisture plot measured against a mixing ratio line yields information about its mixing ratio, i.e. the amount of water vapor it contains. A temperature plot measured against this line yields saturation mixing ratio, i.e. if the mixing ratio rises to this value then the parcel is considered saturated. In surface chart analysis, dewpoint is often used to assess the amount of moisture, but on a sounding, mixing ratio serves this function and unlike dewpoint is a conserved value.

### 3.2.6. Thickness scales

Some sounding diagrams, especially traditional paper ones, will have a thickness scale overlaid on them. The forecaster draws a line connecting the 500 mb temperature value with the 1000 mb temperature value, and reads the 1000-500 mb thickness value from this scale. Sounding charts may contain thickness scales for 700-500 mb[1] and other layers as needed.

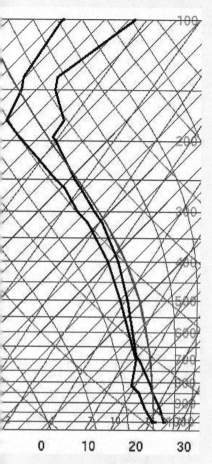

**Figure 3-4. Sounding taken during rain.** This was obtained from Washington Dulles Airport, Virginia, in the evening in July. Light continuous rain was being reported with bases varying from 4500 to 8000 feet. Key signatures are the coarse featureless appearance to the profiles, a near-saturated moist adiabatic profile, and rain being reported on surface observations. A sharp forecaster will always check the raw METAR observations to gauge the type of weather conditions occurring at the sounding site.

## 3.3. Meteorological data

There are two primary pieces of data drawn on the sounding chart: *temperature* and *dewpoint*. These are drawn vertically on the sounding for all levels sampled by the radiosonde, forming two zig-zag lines. The *temperature profile* is always the rightmost line, and by convention it is drawn in red. It establishes the ELR, or environmental lapse rate. The *dewpoint profile* is the leftmost of the pair, and is usually drawn in green.

When plotting data and constructing lines on the sounding, it's extremely important to not do so blindly. Always be conscious of what the lines being drawn represent, and if they're not understood, seek them out in this book or elsewhere. Some lines will represent actual data, while others, such as those used to construct the LCL, are used only as guidelines. In this book we will try to distinguish the two.

### 3.3.1. Spotting bad data

Humid conditions in the stratosphere are an indicator of radiosonde sensor problems, since the relative humidity in the stratosphere is almost always close to 10%. Radiosonde sensors are not very accurate at these extreme pressures and temperatures, but suspected faulty sensor issues can sometimes be evaluted by checking for excessive moisture profiles in the stratosphere. Expanded 1000-10 mb charts are available at weather.uwyo.edu/upperair.

### 3.3.2. Relative humidity

Relative humidity is commonly described as the ratio of the amount of moisture to the amount the atmosphere can "hold", though

Relative humidity can be easily computed at any level by obtaining the mixing ratio, $e$, (from the environmental dewpoint line) and the saturation mixing ratio, $e_s$, (from the environmental temperature line). It is then computed as RH=$100(e/e_s)$. It can also be estimated by observing the horizontal separation between these dewpoint and temperature lines; the greater the separation, the greater the dewpoint depression and the lower the relative humidity.

### 3.3.3. Standard atmosphere

A fixed model of atmosphere variables through a column based on is referred to as a *standard atmosphere*. Most importantly this provides a distribution of temperature for various altitudes.

There are several different types. One is the ISA (International Standard Atmosphere), developed and published by the International Organization for Standardization (ISO). It

**Figure 3-5. Stability and instability**, as expressed in a cartoon. When the environment is stable, any displacement results in the mass returning to its original location. The velocity magnitude diminishes as the mass seeks its position of equilibrium. Here, the equilibrium position is the bottom of the hole. However if there is instability, any displacement produces an acceleration and increasing velocity. *(US Air Force)*

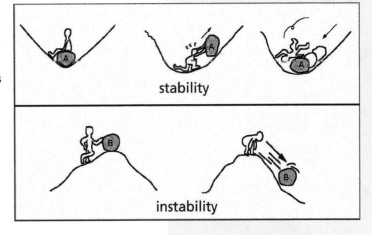

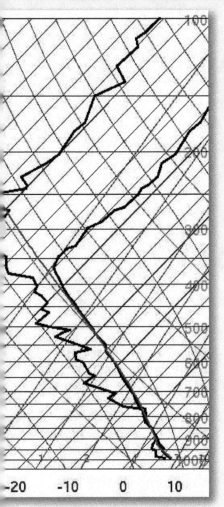

**Figure 3-6. Cold unstable polar air mass sounding** observed in Washington State in November. Key signatures are the very low tropopause (375 mb, or about 24,000 ft), cold column with –33°C at 500 mb, a moist adiabatic lapse rate, and saturated low level conditions. This sounding is associated with a showery, rainy pattern, gusty winds, and icing due to a deep saturated subfreezing layer (4000 to 8000 ft MSL).

Another standard atmosphere is the ICAO Standard atmosphere, and the U.S. Standard Atmosphere (1976).

## 3.4. Static stability and lapse rate

Static stability describes the motion of a parcel displaced vertically. The major factor contributing to the stability of a layer is its *lapse rate*. This is the rate of temperature decrease with height. A high lapse rate value means strong cooling with height. Static stability is exceptionally important to forecasters, as instability favors the ascent and descent of parcels, causing weather, while a statically stable layer will resist such motions.

The basic lapse rate is described as $d\theta/dz$, or the change in potential temperature with height as described by the environmental temperature profile. If it is positive, potential temperature is increasing with height. It is relatively warm at the top of this layer, so parcel ascent along the dry adiabats on the skew T diagram shows that the parcel will be cooler than the surrounding air as it rises. Being negatively buoyant, it will return to its original level. It resists displacement.

Likewise, a negative $d\theta/dz$ in the layer indicates a decrease in potential temperature with height, meaning very rapid cooling with height. The layer is statically unstable, and vertical motion accelerates further. In synoptic and mesoscale meteorology, such a layer does not exist because the air comprising the layer itself will spontaneously rise, overturning the layer and mixing out immediately. If layer $d\theta/dz$ is zero, then the layer is statically neutral.

Layers with steep lapse rates are frequently associated with severe weather. A lapse rate of 6.7° C/km has been found to be a useful lower limit for tornado outbreaks (Craven 2000). Severe weather indices like the Showalter Index and the lifted index indirectly measure this lapse rate, providing a value that can be assessed for severe weather potential.

The Vertical Totals (VT) index (q.v.) is an index comprised entirely of the lapse rate between 850 and 500 mb. Its use in higher terrain is questionable because the 850 mb is so close to the ground, but at low elevations (under approximately 1000 ft ML) it still remains a very simple and useful parameter. Steepening mid-level lapse rates are often associated with upper-level lift and may indicate the approach of a

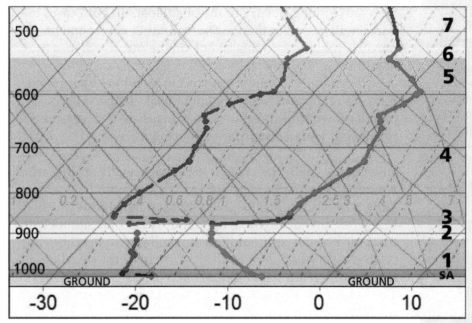

**Figure 3-7. Temperature profile** (red) and moisture profile (green) at Longview, Texas on January 21, 1985 at 0000 UTC in the wake of a severe cold front. The lowest layer, 1, shows conditional instability. Intense modification over the warm ground is causing a thin superadiabatic layer, SA, about 900 ft deep. It is being heated faster than turbulence can dissipate the gradient. This is adding instability to the air mass from the bottom, producing a dry adiabatic lapse rate in 1. Layer 2 shows a relaxed temperature gradient and a stable layer. Layer 3 shows a very strong inversion associated with the front. Temperatures here rise from -18 to -11°C in only 290 feet. Layer 4 shows a continuation of a strong, stable inversion but at a relaxed lapse rate. This is part of the cold air mass. Note how the dewpoint trace shows rapidly rising dewpoint, a common characteristic of frontal inversions. The very top of this layer (4) is the frontal surface aloft, at 14,000 ft MSL. In layer 5 we pick up the tropical air mass, which is conditionally unstable. Layer 6 shows a weak inversion of unknown origin, and layer 7 shows more of the moist tropical air.

quasi-geostrophic disturbance, while weakening lapse rates indicate upper-level subsidence.

### 3.4.1. Static stability

At a height of about 10,000 ft and at typical springtime temperatures, the dry adiabatic lapse rate ($d\theta/dz$) is 9.8 °C/km, while the moist adiabatic lapse rate ($d\theta_e/dz$) is 5.3 °C/km. This is the amount of cooling that a rising parcel would experience when it is unsaturated, and saturated, respectively. The much slower cooling rate in saturated conditions illustrates the contribution of latent heat. This latent heat is caused by the phase change of water vapor to liquid. The change from 9.8 to 5.3° C/km shows that 4.5 °C/km of latent heat has been added to the parcel.

If the lapse rate of a layer is less than both the dry adiabatic and moist adiabatic lapse rate, it is said to have static stability. In other words, both $d\theta_e/dz$ and $d\theta/dz$ are positive. This is described by the lapse rate relationship $\Gamma_e < \Gamma_m < \Gamma_d$, where each variable represents the environmental lapse rate, the moist adiabatic lapse rate, and the dry adiabatic lapse rate, respectively. It shows that the environmental lapse rate $\Gamma_e$ is weaker than both the moist and dry adiabatic lapse rate.

A parcel can only ascend along the dry adiabats and moist adiabats. If it begins an ascent within this layer, it will start with a point that is part of the temperature profile and

**Pseudo-adiabatic diagrams**
All diagrams used in meteorology are considered to be pseudo-adiabatic. This means that air parcels are heated by latent heat and any condensed moisture exits the parcel.

then move along the dry or moist adiabat. In both cases the parcel will be cooler than the surrounding environment. If a parcel is moving into this layer from below, it will lose momentum.

Even under optimal circumstances where a saturated parcel is rising and receiving latent heat, the parcel is cooler than the surrounding environment, causing it to be negatively buoyant with a downward restoring force.

**Inversions.** Stable layers that are exceptionally strong, where temperature remains constant or increases with height, are known as *inversions*. However the term may loosely be used to refer to any layer where temperatures depart significantly from the normal configuration of cooling with height. The AMS glossary defines it as "departure from the usual decrease or increase with altitude" of a property (AMS, 2017). The base of the inversion is the lowest temperature at which the inversion occurs.

**Subsidence inversion.** The subsidence inversion is a layer aloft which shows dry air within the inversion. It is caused by the sinking of air aloft, often by a quasigeostrophic disturbance acting on the middle or upper troposphere.

**Radiational inversion.** The radiational inversion is based at the surface and is caused by nighttime heating. It is identified by an inversion based at the surface during the night, and in which dewpoint depression increases with height.

**Frontal inversion.** The frontal inversion is distinctive because dewpoint depression is the same or decreases through the inversion. In other words, the dewpoint line tends to move parallel with the temperature line. Most of the time the frontal inversion is surface-based, but elevated frontal inversions can sometimes occur.

### 3.4.2. Conditional instability

If the lapse rate of a layer is greater than the moist adiabatic lapse rate but less than the dry adiabatic lapse rate, it is described as conditionally unstable. This is described by the lapse rate relationship $\Gamma_m < \Gamma_e < \Gamma_d$, showing the environmental lapse rate $\Gamma_e$ falls in between the dry and moist adiabatic lapse rates. The term $d\theta_e/dz$ is negative but $d\theta/dz$ is positive. This is the typical state of large portions of the troposphere before severe weather events.

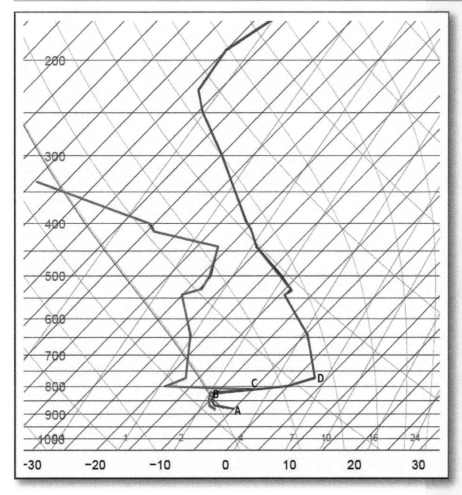

Figure 3-8. **Skew-T diagram showing a strong inversion** for Salt Lake City during a multiple-week plateau high event with trapping of fog in the valleys. The fog is reinforced with nighttime radiation of the cloud tops in the clear air. The fog settled in on December 16, 1985, with a temperature steady between 19°F and 27°F, and did not clear up until January 1, 1986.

Kinetic energy in a conditionally unstable atmosphere is released whenever a rising parcel is saturated and latent heat is being added to the parcel, causing it to rise at the moist adiabatic lapse rate. However if the layer is dry, any forced lift will cause the parcel to lift at the dry adiabatic lapse rate, causing negative buoyancy, and the parcel will seek out its original level.

Environments with steep mid-level lapse rates are associated with conditionally unstable layers. The 850 to 500 mb lapse rate, expressed as the vertical totals (VT) index, are most useful for identifying these layers.

Conditionally unstable layers are very common, especially during the warm season. If sufficient moisture is available in the lower part of the atmosphere, showers and thunderstorms may develop.

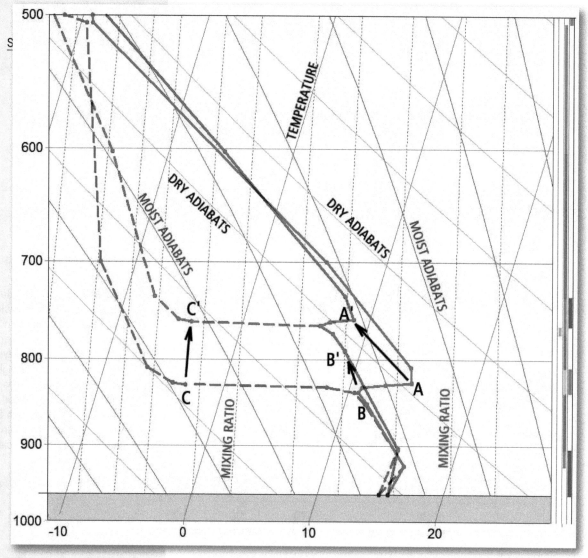

**Figure 3-9. Example of potential instability.** Here we have lifted the column roughly 100 mb from its previous state (blue) to the new state (green dewpoint and red temperature). The sounding had previously showed a "loaded gun" configuration with a cap. The capping inversion, which is dry, moved dry adiabatically from A to A'. However the moist boundary layer saturated and moved moist adiabatically from B to B'. The result is the erosion of the cap. Due to the formation of a MAUL

### 3.4.3. Absolute instability

If the lapse rate of a layer is greater than the dry adiabatic lapse rate, the layer is described as absolutely unstable. Both $d\theta_e/dz$ and $d\theta/dz$ are negative. Because this lapse rate exceeds both the dry and moist adiabatic lapse rates, this lapse rate is described as superadiabatic. This is described by the lapse rate relationship $\Gamma_m < \Gamma_d < \Gamma_e$, showing the environmental lapse rate $\Gamma_e$ is larger than both the moist and dry adiabatic lapse rates.

Superadiabatic lapse rates will cause spontaneous overturning and mixing of the layer, so these lapse rates are almost never encountered on a sounding. Intense heating near the ground, however, can produce very shallow layers of surface-based superadiabatic air, and there is evidence that transitory mid-level superadiabatic layers can exist on small

**Figure 3-10. Conditional symmetric instability** developing over Oklahoma in December 2010. We see the appearance of nimbostratus clouds with a diffuse base, producing snow.

time and space scales, which are replaced by temperature advection or diabatic processes faster than the atmosphere can equalize the layer.

Superadiabatic lapse rates may appear on soundings due to radiosonde error. A common source of this is evaporational cooling of the sensor as the radiosonde ascends from the top of a cloud with high liquid content into clear air. If such errors are not identified and corrected, they will interfere with stability index values derived from the sounding.

### 3.4.4. Potential instability

When a *layer* is acted on by dynamic lift, this can produce stability changes over time. This is known as potential instability, sometimes referred to as "convective instability".

For example, we will consider the entire troposphere as a layer. The lower half of the layer consists of saturated air and the upper half consists of dry air. Lifting the entire sounding by 1000 meters will cause latent heat release of the lower portion as parcels at various levels follow the moist adiabatic lapse rate. Meanwhile in the upper half of the layer, parcels lift at the dry adiabatic lapse rate, undergoing rapid cooling with height. Overall, the resulting state at a later time shows a much steeper environmental lapse rate

**Isolated storm coverage**
Low cell counts may be caused by:
* Weak lapse rates / weak CAPE
* Weak moisture advection
* A lack of large-scale ascent
* Strong capping (high CINH)

**Extensive storm coverage**
Numerous cells may be caused by:
* Steep lapse rates / high CAPE
* Strong moisture advection
* Large-scale ascent
* Weak capping (low CINH)

The processes that result in weak storm coverage also tend to produce an absence of storms; i.e. the "bust day" or "busted forecast".

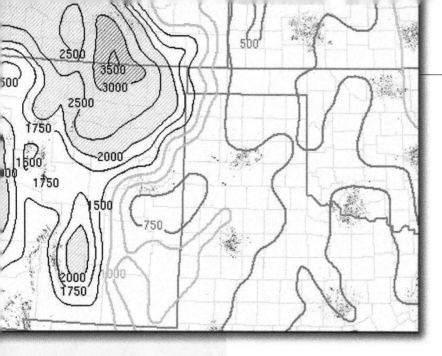

Figure 3-11. **Lifted condensation analysis** as obtained from the SPC Mesoscale Analysis. The source of this data is RAP model output fields. *(NOAA)*

through the entire layer. Therefore the layer at the start of this process is said to contain potential instability.

Potential stability/instability is a function of both temperature and moisture at all levels. Therefore, we define potential instability to exist when either wet-bulb potential temperature ($\theta_w$) or equivalent potential temperature ($\theta_e$) decreases with altitude. If these properties increase with altitude, then the layer shows potential stability. If these properties are the same throughout the layer, it is said to be potentially neutral.

### 3.4.5. Symmetric instability

Symmetric instability refers to acceleration which can occur in an environment which is inertially and statically stable, where parcels tend to return to their original position both horizontally and vertically. It is assumed that parcels tend to "cling" to their surface of potential temperature, $\theta$, and momentum, $M$.

In a typical weather pattern, the potential temperature, or isentropic, surface lies relatively horizontal, while the momentum surfaces tend to be vertical. A forced upward displacement causes the parcel to move where potential temperatures are warmer, so the parcel sinks back to its original level because it is relatively cool. A similar process occurs with momentum when the parcel is forced horizontally. Both of the *restoring forces* cause the parcel to move back to their original level.

If the slope of the momentum surface is brought horizontal and the potential temperature surface more vertical so

---

**Cloud height**

The formula for estimating cloud height based on surface data (such as METAR data)

$$z_{LCL} = 222 \times (T - T_d)$$

where T and Td are the temperature and dewpoint, respectively, in degrees Fahrenheit, and $z_{LCL}$ is the cloud base height in feet, which provides a useful estimate of the lifted condensation level (LCL).

its slope exceeds that of the momentum surface, this causes an important change where the restoring forces act away from the parcel's original displacement.

CSI is normally found during the cool season, with a moist adiabatic lapse rate (weak instability or slightly stable), and in a warm frontal zone, often in southwesterly flow. Convection often takes on a banded appearance. Winds aloft tend to be strong at and above the convection level, with unidirectional flow. Precipitation does not involve cumulonimbus clouds or thunder, as these are associated with convective instability.

### 3.4.6. Moist absolute instability

Moist absolutely instability (Bryan and Fritsch 2000) is similar to conditional instability but it is defined by a layer with very low dewpoint depression, less than 1°C, and whose lapse rate exceeds the moist adiabatic lapse rate. It is often a mesoscale process, with horizontal scales in the tens of km and lifecycles of over 30 minutes, and can span over 100 mb of depth. Layers with moist absolute instability are referred to as a moist absolutely unstable layer (MAUL). It is caused by the forced lifting of a saturated layer and maintenance by mesoscale processes that keeps it from dissipating.

These are noteworthy because they are inherently unstable: any displacement will cause continued motion. It would seem that MAUL layers would immediately overturn and mix out, but there is evidence that meteorological processes can help sustain the MAUL faster than it can equalize.

The existence of a MAUL should not be mistaken for errors originating from the radiosonde. Evaporative cooling of a wet sensor emerging out of the top of a cloud may produce artificially steep lapse rates.

## 3.5. Lifting

Layers comprising a sounding profile are made up of either stable or conditionally unstable layers. The only way that a parcel will have immediate buoyancy is if it is saturated and in a conditionally unstable layer. Therefore, a lifting process is normally necessary for a parcel to reach its LCL where latent heat can be added to the process and it can rise moist adiabatically. Once it reaches its LCL, it is not neces-

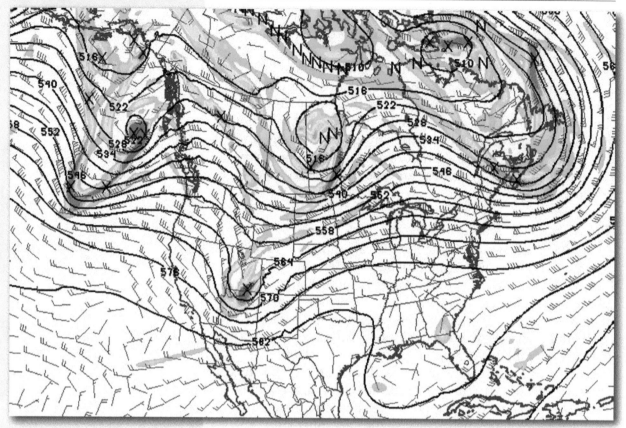

**Figure X-XX. The 500 mb height and vorticity chart** is commonly used to look for areas of large-scale dynamic lift. This affects a considerable depth of the troposphere and causes a surface response: pressure falls and cyclogenesis. This lift also destabilizes the column and increases lapse rate. Lift is generally found downstream from the high cyclonic vorticity areas, such as the major short wave trough seen here over northern Arizona. This implies that strong upper level lift is affecting the Four Corners area.

sarily buoyant; it may still be cooler than the surrounding air, and continued lift is necessary.

### 3.5.1. Dynamic lift

This occurs when a quasi-geostrophic disturbance, like a short wave trough, is embedded in the environmental flow and approaches the target area. It lifts the column of air, either some of it or most of it, causing adiabatic cooling across a broad region. This steepens the column and by doing this it can directly destabilize the air mass and remove capping.

### 3.5.2. Isentropic lift

Isentropes, or lines of equal potential temperature, form surfaces whose value always increases with height except in an absolutely unstable atmosphere. The height is proportional to the density in the lower layers and especially to deep layers of dense air. This means that isentropic surfaces are displaced upward over cold air masses and are lower in warm air masses.

If the wind pattern is oriented so that flow "climbs" upward over a region of high isentropic surfaces, isentropic lift is observed. This naturally tends to happen along warm fronts and explains how frontal ascent occurs. Parcels may saturate and produce clouds and precipitation if the relative humidity and amount of lifting is sufficiently high. Once clouds and precipitation are generated, or *any type of diabatic heating or cooling occurs*, parcels will no longer move isentropically.

### 3.5.3. Mass convergence

Lift can be achieved through low-level forcing. This usually occurs through convergence along fronts and drylines, and can be augmented by thermal circulations.

Not all drylines and fronts have strong convergence. In some cases, streamlines are parallel to a dryline, with south winds on both sides of the boundary. In other situations, there is very strong convergence. This type of convergence is often related to the strength of upper-level divergence and the resulting pressure falls. If convergence is weak, this can be a significant favor in maintaining convective inhibition.

### 3.5.4. Differential warming

The air behind the dryline, or in any strongly heated area, creates an area of upward motion, compared to the sinking which dominates cool areas, and this produces a solenoidal circulation that flows toward the heating. In the case of the dryline, the solenoidal circulation flows across the dryline toward the west, meeting up with the west-to-east flow in the warm dry air, intensifying the dewpoint gradient and producing upward motion. In short, areas of enhanced lift in the lower levels can be driven by daytime heating.

### 3.5.5. Orographic effects

Air may lift simply by moving upslope or up hills and mountains and being forced to rise in the process. Local topography may also create directional differences and create areas of low-level divergence and convergence. This can be a significant source of localized upward motion. Local features can be noteworthy sources of orographic lift; such as (in the Great Plains) the Texas Caprock, Palo Duro Canyon, and the Palmer Divide in Colorado.

## 3.6. Basic parcel properties

Provided here is information on parcel properties, similar to Chapter 1, except here information is provided on calculating them with the skew-T log-p diagram.

### 3.6.1. Potential temperature

Potential temperature ($\theta$) is the temperature a parcel would have if it was forced to 1000 mb, which is near 360 ft MSL in a standard atmosphere. It provides commonality with temperatures at other stations at different elevations and can be extremely useful for frontal analysis, particularly in mountainous areas.

To compute it on the skew-T diagram for a given environmental temperature $T$ at a given level $P$, potential temperature can be determined by simply reading the dry adiabat that passes through $T$, or bringing a line from $T$ down the dry adiabat to the 1000 mb level, and reading the temperature.

### 3.6.2. Lifted condensation level

Lifted condensation level (LCL) (Saucier 1955) is simply the level at which an air parcel rising dry air becomes saturated. At this level it has cooled adiabatically to its dewpoint for that level. The rising parcel is comprised of either dynamically lifted parcels or rising thermals.

The LCL is found by selecting an initial temperature and dewpoint at a specific level, usually the surface, and the temperature is lifted along the dry adiabat until it reaches the mixing ratio line which intersects the initial dewpoint. This represents the parcel cooling until (mixing ratio).

A common assumption is that the parcel cools until it reaches its dewpoint temperature. However mixing ratio, and not dewpoint, is conserved as the parcel rises and pressure changes. The dewpoint actually decreases slowly as the parcel rises, at a rate of roughly 1 °C per 2000 ft.

In operational meteorology when locating a forecast target area, there are various sources for LCL data including radiosonde displays, the SPC Mesoscale Analysis, and HRRR output fields. It is important to be aware of the sources of data and their limitations. Radiosonde data uses the actual radiosodne report but is only available at specific times. The SPC Mesoscale Analysis and HRRR output are modeled fields, they are at least an hour or two old, and they are not necessarily using all available stations.

Surface data and the equation for cloud height using temperature may provide an estimate of the LCL. A radiosonde observation however is the definitive measurement of LCL.

### 3.6.3. Convective condensation level (CCL)

Normally soundings are only available during the nighttime or morning hours. A surface-based parcel will be much cooler than the environment if listed dry adiabatically. If the dominant weather process during the next 12 hours will be surface heating, forecasters should compute the convective condensation level (CCL). By definition, this is the lifted condensation level given by a surface-based parcel that has heated sufficiently for parcels to overcome all negative buoyancy and produce deep convection. This surface temperature is the *convective temperature* ($T_c$).

The CCL is constructed by drawing a line from the surface mixing ratio value up the mixing ratio line to the environmental temperature line. This line represents the new environmental dewpoint profile at the time deep convection begins. The isobar (pressure line) where this intersection occurs is the CCL level. From the CCL, a line is brought to the surface along the dry adiabats, representing the new temperature profile at the time deep convection is presumed to begin.

Forecasters are free to make adjustments to this technique. Most of the time, thunderstorms form somewhere in a forecast area several degrees before the $T_c$ is reached, so it may be more helpful to use a lower CCL and a cooler dry adiabat.

If the convective temperature exceeds the forecast maximum temperature, then the environmental temperature aloft is probably too warm or too "capped", and showers and thunderstorms are not likely.

If the convective temperature is significantly below the forecast maximum temperature, this is a sign that thunderstorms will form during the morning hours or the early afternoon, with widespread coverage. Severe weather potential may be reduced due to the widespread cell competition.

**Differential thermal advection**

Forecasters should avoid using fifferential thermal advection to predict destabilization in the middle and upper levels (above 850 mb). Although it may seem that cold air advection approaching the target would cause cooling over time, destabilizing the air mass, this does not take into account thickness changes, diabatic processes, and other factors that shape the upper-level height field. Always use model forecast data where available to look for trends in layer destabilization. This can be ignored at lower levels (850 mb, 925 mb, and surface) where there should be a strong relationship to the isobars/contours and the isotherms.

### 3.6.4. Level of free convection (LFC)

The level of free convection is the level at which negative buoyancy changes to positive buoyancy, as a result of a parcel ascending past a mid-level capping inversion. It is computed based on a specific parcel lift method.

A low LFC is associated with a low cap, and possibly greater CAPE than normal. It is also associated with high NCAPE and high CAPE density.

An LFC may still be calculated even if deep convection is impossible, because the parcel's adiabatic temperature changes are known all the way to the top of the diagram. Therefore, before using an LFC, forecasters must also answer the question of whether deep convection is likely.

### 3.6.5. Equivalent temperature ($T_e$)

The purpose of equivalent temperature is to extract the maximum latent heat in the parcel, referenced to the parcel's original level. It rises with increasing water vapor (dewpoint), and rises less so with air temperature.

For a given set of environmental $T$ and $T_d$ at a given level $P$, potential temperature can be determined by calculating the lifted condensation level for $T$ and $T_d$ starting from $P$, and lifting from $T$ along the dry adiabat to the LCL. However if the parcel is already saturated, then simply start at T. Then lift from this point along the moist adiabats to the very top of the chart, then return down the dry adiabats to the original level $P$. Then read the temperature $T_e$ at this level.

### 3.6.6. Equivalent potential temperature

Equivalent potential temperature ($\theta_e$) extracts the maximum possible latent heat from the air parcel, similar to equivalent temperature, but it references the final temperature to the 1000 mb level. To compute it on the skew-T, using observed $T$ and $T_d$ at a specific level $P$, first lift the parcel to the top of the diagram. The parcel is lifted along the moist adiabat when reaching the LCL. Then force it down to 1000 mb dry adiabatically. Read the equivalent potential temperature $\theta_e$ at this level, converting it to Kelvin.

This is slightly different from equivalent temperature in that the parcel descends to 1000 mb, not to its original level. The process of moving it to 1000 mb does not represent an adiabatic process, but is rather to obtain the temperature with reference to a common level.

---

**Virtual temperature**

A good way to remember whether moist or dry air is denser is to consider the constituent molecules. A water molecule, $H_2O$, contains two atoms of hydrogen and one molecule of oxygen. Hydrogen is the lightest element in the periodic table, whereas dry air is made up mostly of nitrogen, carbon, and oxygen. Therefore the more water vapor is added to a parcel, the less dense it is.

Virtual temperature is the temperature that dry air, with all moisture removed, must have in order to have the same density. The temperature must increase, reducing the density. The more moisture there is, the warmer the virtual temperature must be.

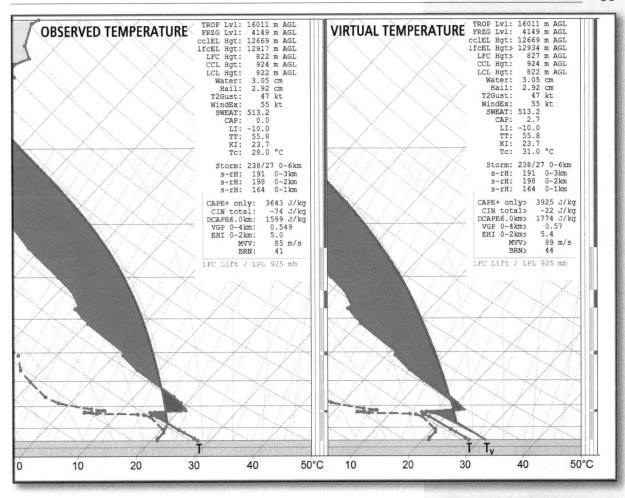

**Figure 3-12. The use of virtual temperature.** This diagram is for May 24, 2011, when EF4 tornadoes struck west of Oklahoma City. This is the Norman sounding for 1800 UTC, modified with the afternoon maximum temperature, 28°C. This shows standard air temperature parcel lift (left) versus virtual temperature (right), where the purple line indicates virtual temperature. The LCL does is based on air temperature, not virtual temperature, so the LCL height is the same as that for the standard parcel lift. Note that the moist air shifts the virtual temperature well to the right. The negative area is significantly weaker, and CIN drops from 74 to 22 J kg$^{-1}$, a more representative value for initiation.

This value, theta-e, is used frequently in operational meteorology as this expresses a measure of warmth of updraft parcels within deep convection. Given equal upper air condictions across a region, the area with highest theta-e will have the most buoyant updrafts, the highest instability, and the most severe weather. Note that when we calculate this value, the parcel does not literally return to 1000 mb; instead this is meant to use 1000 mb as a common level to compare other theta-e readings in the area regardless of altitude.

### 3.6.7. Saturated equivalent potential temperature

Saturated equivalent potential temperature ($\theta_{es}$). For a given set of environmental $T$ and $T_d$ at a given level $P$, rise along a moist adiabat from T to the top of the chart, then return it along the dry adiabats to 1000 mb. This shows the equivalent potential temperature that would exist if the par-

> **Estimating virtual temperature**
>
> If making computations by hand, virtual temperature can be estimated with the formula $T_v = T+(w/5.8)$ for Celsius, or $T_v = T+(w/3.2)$ 3.2 for Fahrenheit. For example, if the temperature is 81°F and the dewpoint is 70°F, reading the mixing ratio overprint shows that the mixing ratio is 17 g/kg. Solving this we get a virtual temperature correction of 5°F, which is added to the temperature to give us 86°F. This is very close to the actual number.

cel was already saturated. This is not used often in operational meteorology.

### 3.6.8. Wet bulb temperature ($T_w$)

Wet-bulb temperature expresses the temperature in the parcel after the maximum possible evaporative cooling. For a given set of environmental $T$ and $T_d$ at a given level $P$, calculate the LCL. This is done by rising from $T$ along the dry adiabat until it intersects the mixing ratio for $T_d$, which gives the LCL. Then go downward from the LCL along the moist adiabats, returning to the original level $P$. The temperature at this point gives the wet bulb temperature, usually expressed in °C.

In operational meteorology, wet bulb temperature is used extensively in assessing downburst potential. The distribution of wet bulb temperatures through the vertical forms the basis for DCAPE.

### 3.6.9. Wet bulb potential temperature

Wet-bulb potential temperature ($\theta_{ew}$) is the temperature after maximum possible evaporative cooling in the parcel, similar to wet bulb temperature, however it is referenced to a common level, 1000 mb. To calculate it, for a given set of environmental $T$ and $T_d$ at a given level $P$, determine wet bulb temperature, but continute the line past $P$ along the moist adiabat to the 1000 mb level. The temperature value is the wet bulb potential temperature, usually expressed in Kelvin.

In operational meteorology, this is used in a similar manner to evaluate downdraft potential. Note that parcels are not literally descended to 1000 mb, but rather we are using the 1000 mb level as a common reference level to make it possible to compare the values with other wet bulb readings across a region, removing the effects of elevation.

### 3.6.10. Vapor pressure (e)

Vapor pressure is the contribution to station pressure (actual pressure) by the water vapor molecules. To obtain vapor pressure using the skew-T, given environmental dewpoint $T_d$ at a given level $P$, ascend from $T_d$ along an isotherm to the 622 mb isobar. Then read the mixing ratio value. This yields the vapor pressure in mb.

> This is used for storm-relative helicity. We will now differentiate between a couple of effective layer types you might encounter in forecasting.
>
> The *effective inflow layer* is a band where CAPE is greater than 100 J kg$^{-1}$ and CIN is greater than –250 J kg$^{-1}$. This is used for computing effective storm-relative helicity (SRH), and is part of the supercell composite parameter (SCP).
>
> The *effective layer for bulk wind* extends from the effective inflow layer base upward to the equilibrium level using a most-unstable parcel method.

## 3.6.11. Saturation vapor pressure ($e_s$)

To obtain saturated vapor pressure for a given environmental temperature $T$ at a given level $P$, ascend from $T$ along an isotherm to the 622 mb isobar. Then read the mixing ratio value. This yields the saturated vapor pressure in mb.

## 3.7. Proximity sounding

It is normally assumed that the nearest sounding is representative for storms that develop forecast area. For example, the Norman, Oklahoma sounding is often used to predict the development of storms in Tulsa.

One way to solve the single-station radiosonde problem is to choose two radiosonde sites between which the parcel is located and display them simultaneously with different colors. This is referred to as a *spaghetti sounding*.

It is also possible to blend two soundings, creating a *composite sounding*. The RAOB software package is capable of creating composite as well as spaghetti sounding plots. Composite soundings may also be created by hand using a blank sounding and the data from two radiosonde sites.

Some software packages may offer more advanced methods, such as bilinear interpolation or triangulation of data between multiple stations. This is a crude form of weather map objective analysis, and it offers the most accurate proximity sounding possible without the use of statistical and numerical models.

Forecasters must be skilled in modifying the sounding for *temporal changes* expected as the day continues. The 1200 UTC sounding is rarely representative of conditions by afternoon. Some sources for modifying a sounding include the maximum forecast temperature and model forecast data. The sounding can also be modified using *supplemental data*. NEXRAD VAD/VWP wind data as well as profiler data are frequently neglected sources of data for modifying a sound-

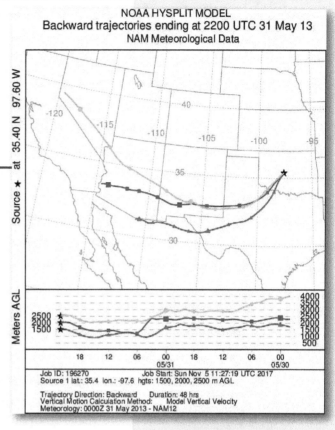

**Figure 3-13. Source of parcels** at 4900 (red), 6500 (blue), and 8000 (green) ft AGL shortly before the May 31, 2013 El Reno tornadic storm. These parcels comprise the air within the cap. The lowest (4900 ft) level is marginally within the cap. This shows that the cap air has its source region in Arizona and New Mexico, and may have actually been brought to these levels through dry convection over the higher terrain.

> **The cap as identified in the 1950s**
> Thunderstorms seem to be damped under the warm "lid" aloft at 700 mb in areas where the greatest advance of isotherms at that level has occurred.
>
> - LYNN L. MEANS, 1952
> U.S. Weather Bureau, Chicago
> *On Thunderstorm Forecasting in the Central United States*

**Figure 3-14 (opposite): Diurnal cycle of the cap.** This shows a cross section of layers associated with the cap (shading) and clouds (gray masses), with time of day (x-axis) and altitude (y-axis). It shows the free atmosphere (FA), the capping inversion (CI), the entrainment zone (EZ), the residual layer (RL), and the surface boundary layer (SBL). Cloud forms include stratocumulus, cumulus, cumulonimbus, and altocumulus. The equals sign identifies WMO cloud type designations. At the top, we see a Skew T log p diagram at 3 to 5 hour intervals, with the skew-T tilted so the isentropes stand up vertically. Line segments tilting slightly to the right are slightly to strongly (conditionally) unstable, and those tilting strongly to the right are stable, It is impossible for segments to tilt to the left as this makes them absolutely unstable. They can exist briefly on small time/space scales in certain situations, and are widespread within the shallow contact layer at the surface with strong heating.

ing, though the modifications are primarily limited to wind data.

Dynamical models can be used to either modify the morning sounding or can be used as the sounding itself. Before the advent of high-resolution mesoscale models, forecast soundings had coarse vertical resolution and important features were often "smoothed out". The 3 km NAM and HRRR offer extensive modeling of cloud physics, precipitation, and radiation processes and have good vertical resolution, and are very useful for forecast soundings.

Forecasters, however, must be attentive to whether the model is representative of actual conditions. It is also important to make sure forecast sounding points are not in areas of convective precipitation being depicted by the model, which will contaminate indices and rules of thumb for that location.

## 3.8. Parcel techniques

One of the biggest challenges in severe weather forecasting is modeling the thunderstorm updraft. This is an important part of using the thermodynamic diagram. Unfortunately this does not have a simple solution, because thunderstorm updrafts ingest air from a layer near the surface with varying depth and composition. The only thing we can be certain of is that an air mass in the lower half of the troposphere is ingested. Typically this is a layer about 1 km deep near the ground, but thunderstorms can also feed from an elevated layer of rich moisture.

There are several techniques in use for representing the lifted parcel. This includes using the surface to represent the updraft, using the most unstable parcel, using a mixed layer, and using an effective layer. We describe them here, with their common abbreviation.

### 3.8.1. Surface based (SB)

The conditions at the surface, either at the current time or at maximum heating, are used to represent the updraft parcel. This is by far the simplest method, however it fails to take into account shallow layers of moisture or varying moisture and temperature aloft. The SB method tends to overestimate instability, particularly in shallow moisture events. It also underestimates instability at night due to the effects of decoupling of the surface layer.

# SKEW-T LOG-P DIAGRAM 87

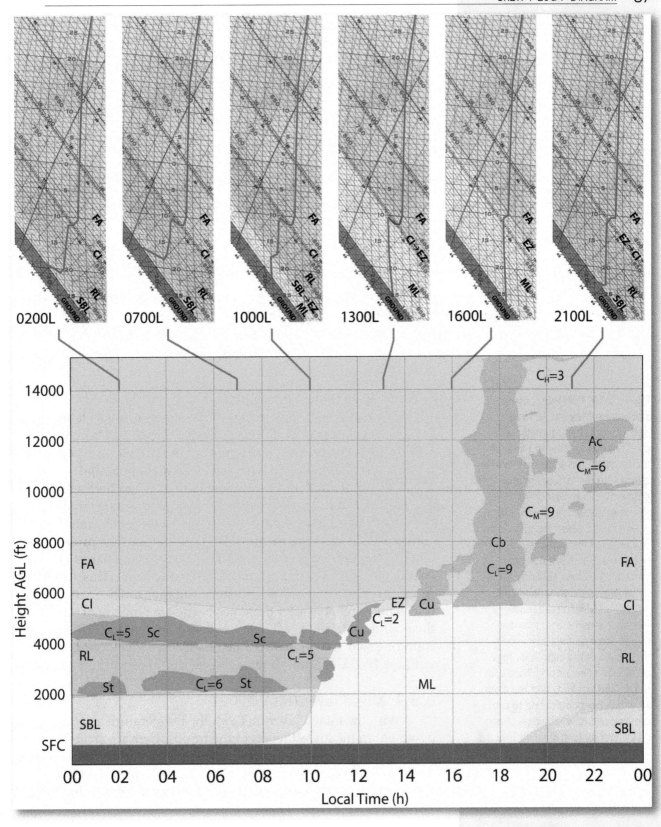

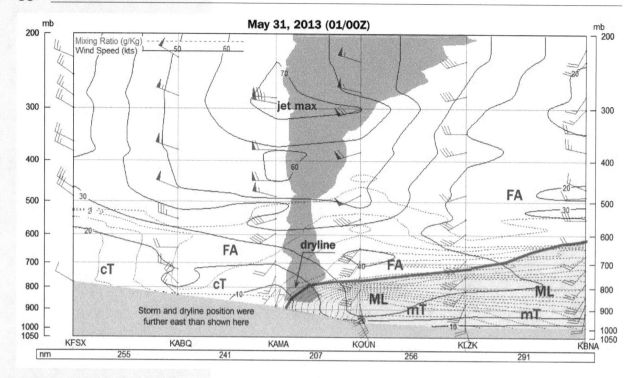

**Figure 3-15. West-east cross section on May 31, 2013**, the date of the El Reno tornado, at 7 pm CDT, shortly after the time of the tornado. It is important to note that due to the lack of upper air data between radiosonde stations, the chart shows the dryline further west than its actual location. This is a problem that forecasters need to be aware of, but it does not diminish the cross section's value in visualizing the synoptic-scale structure. Mixing ratio gradient (visible here) and potential temperature (not shown) shows the outline of the cap. A jet max at 300 mb is in perfect phase with the dryline location, over the storm location after contributing to cap removal and now providing strong bulk shear for long-lived storms.

### 3.8.2. Most unstable (MU)

The most unstable parcel method attempts to locate the parcel in the lower troposphere with the highest wet bulb potential temperature, $\theta_w$ (i.e. furthest to the right with respect to the moist adiabats) or the highest equivalent potential temperature, $\theta_e$, both of these being equivalent for finding the most buoyant air. Typically the lowest 300 mb is evaluated.

The most unstable parcel is highly useful for elevated convection, as it selects the most buoyant parcel. The problem is it frequently overestimates instability, as the parcel is selected on the basis of a reading at only one level. This also causes problems in shallow moisture situations. The method is also more reliable after there has been some mixing of the boundary layer, so MU parcels based on the 1800 UTC sounding will give more accurate results than when used with the 1200 UTC sounding.

### 3.8.3. Mixed layer (ML)

With the mixed layer method, the temperature and dewpoint in the lowest levels of the troposphere are artificially mixed to obtain a mean temperature and dewpoint. To do this, a layer depth is first selected. This is often 100 mb, a value based on experience in the meteorological community.

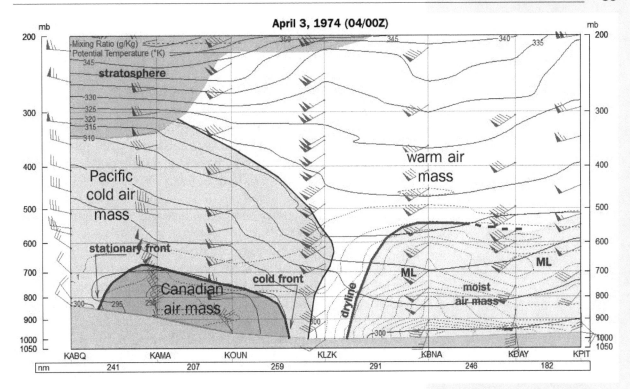

However 150 mb is sometimes used. For the bottom of the layer, the surface is normally used. It can be helpful to mark the top and bottom when working manually with soundings.

The mean dewpoint within this layer is the mixing ratio line that best averages the mean moisture in this layer. Likewise, the environmental temperature profile is mixed along a dry adiabat that best represents the potential temperature in that layer.

This attempts to simulate mixing of the boundary layer due to surface heating, thus the temperature and dewpoint profile can be erased and replaced with these lines to represent complete mixing of this layer, then connected to the rest of the sounding.

The mixed layer method is very reliable for surface-based convection. The 100 mb mixed layer technique has been used in automated analysis schemes at NMC (now NCEP) as far back as the mid-1960s, and still continues to be a favorite scheme for severe weather forecasters. However it will fail in cases of elevated convection, especially instances just north of warm fronts.

**Figure 3-16. West-east cross section on April 3, 1974** (7 pm EST), the date of the supertornado outbreak that produced 148 tornadoes, at 7 pm CDT, shortly after the time of the tornado. The synoptic scale picture was somewhat difficult to sort out from the use of surface data and soundings, but adding the use of cross sections (above) to reconcile features on soundings helped to place boundaries and air masses. This shows the Pacific polar air mass and a deep layer of tropical air over Nashville, close to where tornadoes touched down further east. The dewpoint at Little Rock was only 28°F, and the dryline extended along the Mississippi River near Memphis. Storms formed along boundaries further east within the moist air mass.

> **Drylines**
>
> Since the dryline is not well-recognized outside the regions where it frequently occurs, it is often mistakenly analyzed as a "Pacific front" — i.e., the leading edge of an air mass with a Pacific source region.
>
> CHARLES DOSWELL III, 1982
> *The Operational Meteorology of Convective Weather*

### 3.8.4. Effective layer (EL)

New techniques exist which use an effective layer for storm-relative helicity. The effective layer is the lowest layer in which CAPE exceeds 100 J kg$^{-1}$ and CIN is less than 250 J kg$^{-1}$. So far this has not been applied to CAPE or CIN, although this is possible.

NMC (now NCEP) used an effective layer technique in the mid-1960s (Stackpole 1967), identifying a 100-mb deep mixed layer parcel with the highest wet bulb potential temperature in the lowest 160 mb of the atmosphere. Although the 160 mb depth still fails to detect some elevated instability, this was used to compute lifted index and Showalter index at least until the 2000s.

An effective layer was again proposed for CAPE in a 2014 thesis paper by Adam Cavender (Cavender 2014), using these same constraints, which Cavender is referred to as eCAPE. It was found to be a slightly better discriminator between strong and weak tornadoes. Effective CAPE is was also being used internally in the SPC copy of GEMPAK NSHARP, but as of press time is not displayed on SPC products, and its value is not known (Thompson 2017). It is possible that it may appear on various tools and Internet products in the near future.

### 3.8.5. Positive area shape

Since the early 1990s, the shape of the positive area has been correlated with severe weather. Forecasters found that significant tornadoes often had a "fat" positive area, while nonsevere events had "tall and skinny" positive areas. This quality has been referred to informally as CAPE density. Fat positive areas are associated with large parcel-environment temperature differences just above the LFC, suggesting a concentration of strong acceleration in the middle troposphere and the possibility of enhanced stretching of parcels in the updraft.

One measure of positive area shape that is in common use is NCAPE, or normalized CAPE (q.v.). This is essentially the CAPE magnitude divided by the height of the positive area. High NCAPE values are often associated with severe weather.

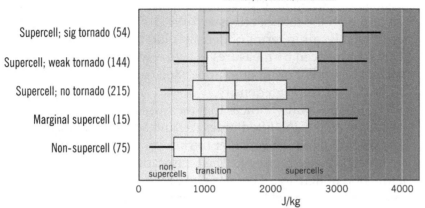

**Figure 3-17. Box and whiskers verification of MLCAPE**, one of the most commonly used predictors of severe weather. This shows skill in forecasting all supercell storms, but no ability to discriminate tornadic supercells from nontornadic cells. *(By author; adapted from Thompson, Edwards, and Hart 2003)*

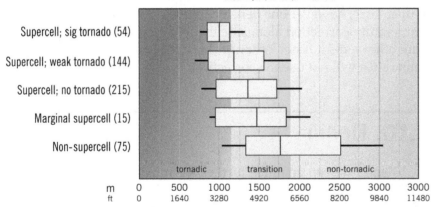

**Figure 3-18. Box and whiskers verification of MLLCL, mixed layer lifted condensation level**, which is often used as one of the predictors for tornadoes. This shows strong skill forecasting significant tornadoes, but a lot of overlap between weak tornado and nontornadic events. This also shows that the parameter must be used with other predictors. It is noteworthy that most non-supercell events were associated with higher LCLs, underscoring the value of high relative humidity environments in producing severe weather. *(By author; adapted from Thompson, Edwards, and Hart 2003)*

> **About CAPE, from its author**
>
> During the 1960s, Frank Ludlam, a leading light at large, took lively interest in severe convection. Although severe traveling hailstorms had been observed for more than a century and in several countries, the application of radar to deep convection was a development area in the late 1950s and 1960s.
>
> Specifically, Ludlam and Keith Browning (his PhD thesis) published a landmark 1962 paper focused on a radar case study of the severe traveling Wokingham storm of July 9, 1959. This paper motivated several PhD studies at Imperial College, including mine, on the mathematical physics and idealized simulation of cumulus convection in shear.
>
> The early 1970s was *the* inaugural period for numerical modeling and dynamics of deep convection, with Imperial College a leading light. In addition, the international 1973 Global Atmosphere Research Programme (GARP) Atlantic Tropical Experiment (GATE) was based in Dakar, W. Africa. Martin and I were aircraft scientists on some NASA Convair 990 flights. (A truly unforgettable experience was an emergency dive from 25,000 ft after hitting severe turbulence!)
>
> Our 1976 paper on tropical squall lines, partly motivated by participation in GATE, quantified the effects of shear on tropical squall lines, and some important physical and dynamical differences from midlatitude systems. CAPE made its way into deep convection parameterizations for large-scale models; firstly, into numerical weather prediction (NWP) models, and in due course into global climate models (GCMs).
>
> MITCH MONCRIEFF, 2017
> father of CAPE with Martin J. Miller

## 3.9. Virtual temperature

Virtual temperature, described in Chapter 1, is occasionally neglected in forecasting. This can have a detrimental influence on the forecast (Doswell 1994). Virtual temperature considers the air in terms of density rather than temperature. This is influenced by measures of actual water vapor content, such as dewpoint or mixing ratio. Moist air has low density, while dry air has high density.

Severe weather indices that measure buoyancy, specifically CAPE, CIN, lifted index, SI, and several others, are measurements which consider the buoyancy of a parcel relative to its environment. The acceleration of the parcel is a function of density. Therefore, to properly evaluate these quantities, the virtual temperature must be used instead of air temperature.

If a forecaster is working with a paper SKEW-T, it is recommended to draw a parallel virtual temperature profile alongside the temperature line. The margin notes show a technique for obtaining virtual temperature. Forecasters can also use the virtual temperature calculator on the NWS El Paso page < https://www.weather.gov/epz/wxcalc_virtual-temperature>.

Certail tools, like SHARPpy use virtual temperature by default. RAOB must be manually configured to use virtual temperature, however one benefit is it will plot the virtual temperature correction alongside the temperature profile. With other tools it is necessary to check the settings. Many online sites will not use the correction.

## 3.10. Dryline

Before we can go much further, we must mention the dryline, a common feature on the Great Plains, particularly during the transition seasons. It has long been associated with severe storms. From the early years of forecasting all the way to the 1990s, the dryline was often mistaken as a Pacific front on analysis charts. Even nowadays, it is sometimes depicted on central analysis maps as a trough, which shows some disregard for their unique structure and circulation.

### 3.10.1. Dryline definition

The dryline is best defined by a strong moisture gradient separating maritime tropical (mT) air which has origin from the Gulf of Mexico and Caribbean Sea from continental tropical (cT) air with origin from northwest Mexico and the southwest United States. The moisture gradient is typically about 10 g kg$^{-1}$ per 100 km, in other words, a dewpoint change from 70° to 40°F over a 60-mile distance.

A dryline may have a wind shift or a thermal gradient, but the dryline is defined by the gradient of dewpoint or mixing ratio and is placed on the moist side of the gradient. Some rules of thumb specify placement at the 55°F isodrosotherm, but this is strongly advised against, as it can cause forecasters to miss the actual location or place it in a location it does not exist.

The dryline is a north-south feature most commonly found in west Texas and western Kansas, eastern New Mexico, and eastern Colorado. On rare occasion with a strong frontal system it may move east of Dallas, Tulsa, and Kansas City, but almost never reaches further east than the Mississippi River.

### 3.10.2. Dryline characteristics

The dryline tends to move eastward into the moist air during the day, and move westward (retrograde) at night. This pattern is called sloshing, and it may last many days or even weeks. The distance it covers during the day is usually about 50 to 200 miles.

As the dryline moves eastward over a given station, it will change the air mass and the temperature and moisture profile at that point. Forecasters must anticipate dryline motion and keep track of trends at stations in the forecast area.

The dryline is often undercut by fronts and air masses, especially in the cool season. Essentially what happens is cold air masses from Pacific or Canadian origin move in and replace the dry air mass, displacing it aloft and eliminating the gradient. As these cold air masses cover the region, high pressure is usually found on surface charts, and it persists until the Gulf of Mexico can re-establish a moisture supply into the Southern Plains.

The dryline is also eliminated during the summer months due to the massive influx of tropical air in response to strong heating in the Desert Southwest. It is found further and further west in mid and late June, and the gradients

---

**The derecho, 1880s**

The derecho, or "straight blow of the prairies", is a powerfully depressing and violently progressing mass of cold air, moving destructively onward in slightly diverging straight lines, in Iowa generally toward the southeast, with its storm-cloud front curving as the storm-lines diverge. The barometer trends upwards and the thermometer falls greatly under the blow of this cold air of the upper strata suddenly striking the ground. The derecho will blow a train of cars from its track, unroof, overturn, and destroy houses; but it does not twist the timbers into splinters and drive these firmly into the hard soil of the prairie.

- DR. GUSTAVUS HEINRICHS
Iowa Weather Service, 1888

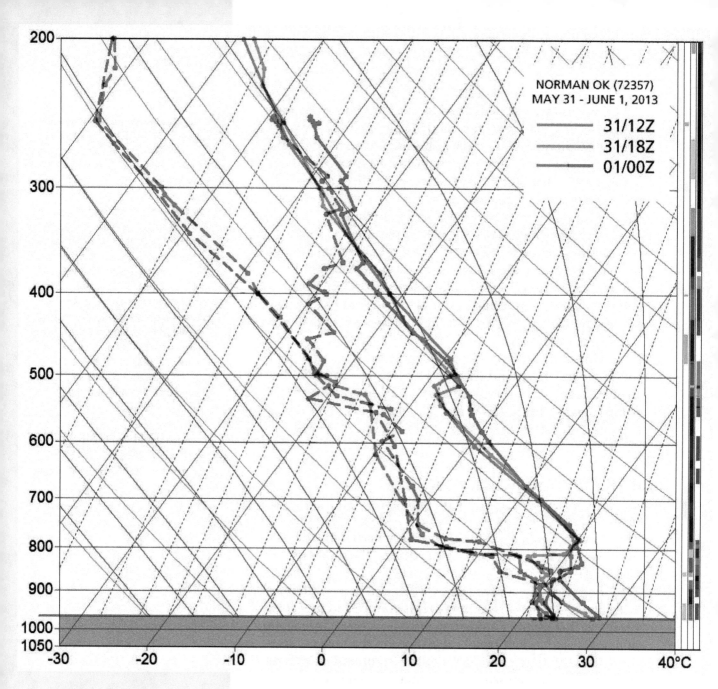

**Figure 3-19. 24-hour changes in the May 31, 2013 El Reno, Oklahoma tornado sounding**, for Norman, Oklahoma (30 miles southeast of the event) progressing from gray (31/0000 UTC, the previous evening); to blue (31/1200 UTC); to orange (31/1800 UTC) and then to the standard red and green (01/0000 UTC). The diagram looks like a confusing jumble of lines, but if we take it layer by layer we can see the changes. In the lowest 900-960 mb layer moisture rapidly increases overnight, then during the day we see surface heating, with last-minute moisture advection bringing the mean layer dewpoint from 21 to 22°C. In the 800-900 mb layer

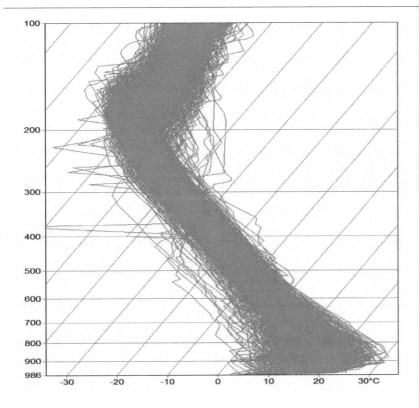

**Figure 3-20. May sounding composite for Norman, Oklahoma**, showing all observed temperatures from 1990 to 2017 at 1200 UTC, around dawn. It reflects a wide variability in temperatures in the lowest 2 km, while most mid-level temperatures fall within a narrow band. The typical tropopause height is about 200 mb (39,000 ft).

become diffuse, eliminating the dryline. It may re-establish itself briefly during the summer but it's only in September and October when heating in the deserts subsides and the dryline becomes re-established in west Texas.

### 3.10.3. Dryline air masses

The dry air mass that is found west of the dryline has its origin in the warm dry plateau area of the desert southwest. It shows steep lapse rates from the surface to the upper troposphere, as strong surface heating is mixed upward. If winds are strong enough and soil moisture is low, it may pick up dust during the day. The dry air mass is part of the same air mass overlying the top of the moisture further east, the elevated mixed layer (EML), so it is strongly related to the capping inversion that affects that region (discussed shortly).

In the moist sector, the air mass has oceanic origin from the Gulf of Mexico, Caribbean, or Atlantic. Sometimes it may consist of recycled polar air from the southeast United States that has warmed over many days, and has mild, moist characteristics. During the cool half of the year, recycled air masses are too dry to support a gradient with the dryline, but during the warm half, dewpoints in the interior regions easily

**The cap in 1927**

Mid-air temperature inversions appear to be quite common [with tornadoes] and the lapse rate next above these inversions very rapid, often nearly or quite of adiabatic value. In short, so far as one can infer from these few observations, the atmosphere in the neighborhood of a tornado appears to be unusually stratified, and tending to become unstable at one or more levels.

W. J. HUMPHREYS
*The Tornado*
Monthly Weather Review, 1927

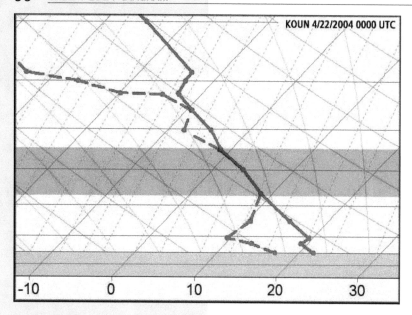

**Figure 3-21. Moist absolutely unstable layer (MAUL)** highlighted in blue. This is identified by a conditionally unstable lapse rate with saturated air (dewpoint depression less than 1°C).

reach the 60°F range due to evapotranspiration, supporting a gradient along the dryline.

### 3.10.4. Dryline lift

The dryline does not produce wind convergence by itself, but winds typically come together along its length. The mass convergence from thsi in the lower atmosphere produces upward motion.

Additionally, differential heating along the dryline is a factor. This produces a solenoidal circulation, with buoyant rising air west of the dryline and sinking air to the east. This drives a low-level component westward, impinging on the west-to-east flow and enhancing the convergence. The solenoidal circulation is enhanced by soil moisture in the moist sector adjacent to the dryline, which increases the amount of insolation required to heat the air mass. It is also be enhanced by sloping terrain, sloping upward under the dryline toward the higher terrain in the dry sector. Both of these are factors that help establish the dryline, often on the Texas Caprock.

## 3.11. Capping inversion

The mid-level inversion known as the *capping inversion*, the *cap*, or the *lid*, has been associated with tornadic storms as far back as the 1920s. This is caused by a layer of warm air in the mid-troposphere, with slightly cooler air below. Beneath the cap, the sounding shows an inversion with static stability or very weak lapse rates with conditional instability. If this occurs in conjunction with rich moisture in the lower levels, a *loaded gun sounding* is the result and severe weather is possible.

The cap typically exists somewhere between 5,000 and 15,000 ft MSL (850 to 600 mb). This warm layer prevents rising parcels from ascending any higher. It is a type of inversion, one which is common in the Great Plains, and it may occasionally be seen in other parts of the United States, Canada, and the world.

### 3.11.1. Cap characteristics

The warm air layer tends to show westerly flow, and has its origins in the southwest deserts of the United States and northern Mexico that has mixed vertically with surface heating and has been advected eastward over a period of days.

The cool air beneath the cap has one of three characteristics:

- stagnant, highly modified polar air with fair weather, mild temperatures, and low dewpoints; or
- recycled polar air with some moisture content and a short trajectory over the coastal Gulf of Mexico; or
- maritime tropical air from deep within the Gulf of Mexico and high dewpoints.

Above the cap is the free atmosphere, also known as the elevated mixed layer (EML), and it is usually has good visibility except when dust is being picked up from the Llano Estacado or western Kansas area, an event that happens on a large scale once every several years. With the presence of a surface dryline, the elevated mixed layer will usually intersect the ground to the west in higher Great Plains terrain, since the air in west Texas or New Mexico during active severe weather outbreaks is, itself, comprised of elevated mixed layer air even at the surface, and shows warm daytime temperatures, cool evenings, strong southwesterly winds, and very low dewpoints.

### 3.11.2. Cap measurement

Prior to the 1980s, forecasters extensively used the 700 mb temperature as a guide. The 10-12°C isotherm was often considered an acceptable starting point for delineating

# 98 SKEW-T LOG-P DIAGRAM

**Figure X-XX. Typical evening dry environment sounding** in the High Plains. This sounding is for autumn in Denver. It shows a low relative humidity environment near the ground, with steep lapse rates from the surface up to about 18,000 ft (500 mb). A cirrus layer with higher relative humidity is found aloft. Note that the surface parcel (red line) shows an absence of CAPE. However mid-level convection can develop in such an environment, and produce very strong downdrafts.

capped from uncapped air (with >10-12° being capped), but this is strongly dependent on the depth and richness of the moist layer, the low-level temperature and elevation, and forcing mechanisms. To overcome these problems, the lid strength index (LSI) was developed in 1980, which related the low-level wet adiabat to that found within the

### 3.11.3. Cap life cycle

The cap begins appearing during the end of polar air outbreak episodes as upper-level ridging moves in and mid- and upper-level temperatures warm up. The lapse rate in the elevated mixed layer is weak or even stable, so even if moisture arrives and cumulonimbus, form they are mostly suppressed, but with enough instability they can form low-topped severe cells.

The cap becomes important when significant upper-level lift approaches. This is most favored when the 500 and 300 mb contours begin to take on an cyclonic appearance across the forecast area. This begins cooling the upper troposphere, increasing the lapse rate. Once the lapse rate becomes conditionally unstable, the stage is set for severe weather and positive CAPE is possible.

The appearance of cyclonic flow and upper-level lift usually means that cyclogenesis is taking place on the high plains. The pressure falls and transverse circulatiosn normal to the jet result in the development of a low-level jet, normally seen at 850 mb but sometimes at 700 mb or 925 mb. This flows northward, initially starting in the Rio Grande Valley of southwest Texas toward west Kansas, and then migrating slightly eastward to the Interstate 35 corridor.

Once the low-level jet and moderate to steep upper-level lapse rates are established, clouds usually flow northward in the residual layer from the previous day's low-level atmosphere. This residual layer extends up to the level of the cap. A radiation inversion often exists at the surface, and if it is strong enough, the surface boundary layer and residual layer decouple, becoming physically unlinked. This allows the low-level jet to accelerate. In many situations with lots of cloudiness the residual layer does not uncouple, and gusty surface winds occur all night at the surface. The formation of a surface inversion is enhanced by a cold surface, which modifies the air mass in contact with the ground. Winds are calm, while stratus races overhead.

Once surface heating commences, the surface layer is quickly eroded. Low clouds in the region west of the low-level jet tend to break up into scattered coversge, while east of the low-level jet, broken to overcast stratocumulus is likely. Beneath these clouds, there is vigorous mixing of the low-level air, producing a layer called a "mixed layer". As mixing continues, the mixed layer temperature equalizes to the same potential temperature, while the dewpoint equalizes to the same mixing ratio.

Within a couple of hundred miles west of the low-level jet, a dryline is often established, and here the cap bends down to make contact with the ground as a dryline. The air west of the dryline is the same as the air above the cap. Storms are most likely to form along the dryline, as usually small-scale upper-level disturbances approach from the west and the dryline is the first location where they come into contact with rich air. The cap is also weaker along the dryline, and the surface winds may show some convergence, helping to weaken the cap further.

As the day progresses, the cumulus fields close to the dryline or the triple point may produce towering cumulus and even brief showers as parcels attempt to penetrate the cap. Some of these towering cumulus may be referred to as "turkey towers" owing to their distinctive shape.

By the time the day ends, storms have either developed and the cap has been broken, or no storms form and for chasers it is called a "bust". After dark, surface cooling results in the re-establishment of a surface boundary layer, and the mixed layer is referred to as a residual layer. The situation unfolds again the next day, modified by the storm events of the previous day and the changes in the surface and upper-level pattern.

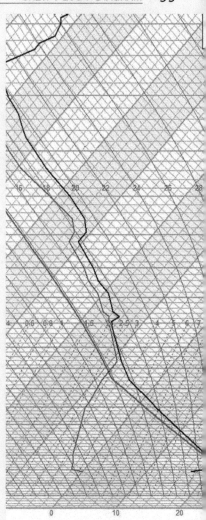

**Figure X-XX. Inverted-V sounding** in the Great Basin area. This sounding is for autumn in Boise. It shows very low relative humidity near the ground, a very steep, nearly dry adiabatic lapse rate from the surface up to about 14,000 ft (600 mb), and moist adiabatic conditions above. This sounding suggests extensive virga, gusty winds, and showers. If there is extensive synoptic scale lift, the column will typically "wet bulb", with strong cooling in the lower levels as evaporation cools these areas to their wet bulb temperature.

## Review Questions

1. What was the earliest type of thermodynamic diagram (1880s)?

2. Describe what is on the x-axis and on the y-axis of the skew-T log-p diagram.

3. How can the moist adiabats be differentiated from the mixing ratio lines on a skew-T log-p diagram?

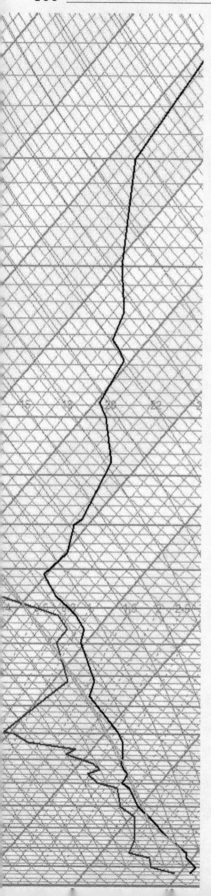

4. Which expressions do theta, theta-w, and theta-e stand for?

5. If one finds the LCL and goes down the moist adiabat to the original level, what value is this?

6. If one finds the LCL and descends the moist adiabat to 1000 mb, what value is this?

7. What is the typical humidity in the stratosphere?

8. What values do e and es represent, and what does the ratio e/es represent?

9. What is the severe weather index that corresponds to 850-500 mb lapse rate?

10. An inversion shows on a sounding where relative humidity is high all the way up to the base of the inversion. What type of inversion is this?

11. The lapse rate of an observed (environmental) layer Γe is given as: Γe < Γm < Γd. What type of stability describes this layer?

12. In a rainy situation an absolutely unstable layer is spotted near the top of a cloud layer. What does this signify?

13. What type of instability is found north of a warm front during the cold season, giving clouds a banded appearance?

14. Potential temperature is referenced to which level? What is the approximate altitude?

15. What is the expression that describes the condensation level of a parcel that has overcome all negative buoyancy?

16. What is the expression for the contribution of the pressure by water in the gaseous state to the total atmospheric pressure? What is the symbol?

17. Which parcel method uses the highest wet bulb potential temperature (wet bulb furthest to the right with respect to moist adiabats)? What is its abbreviation?

18. Which is less dense, moist or dry air?

19. What is the field used to identify the dryline position on surface charts?

20. Once the cap and low-level jet is established, which level is the low-level jet located in?

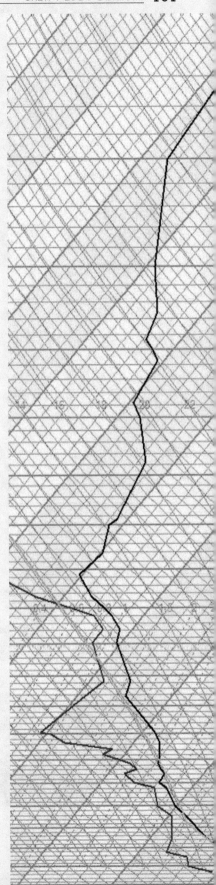

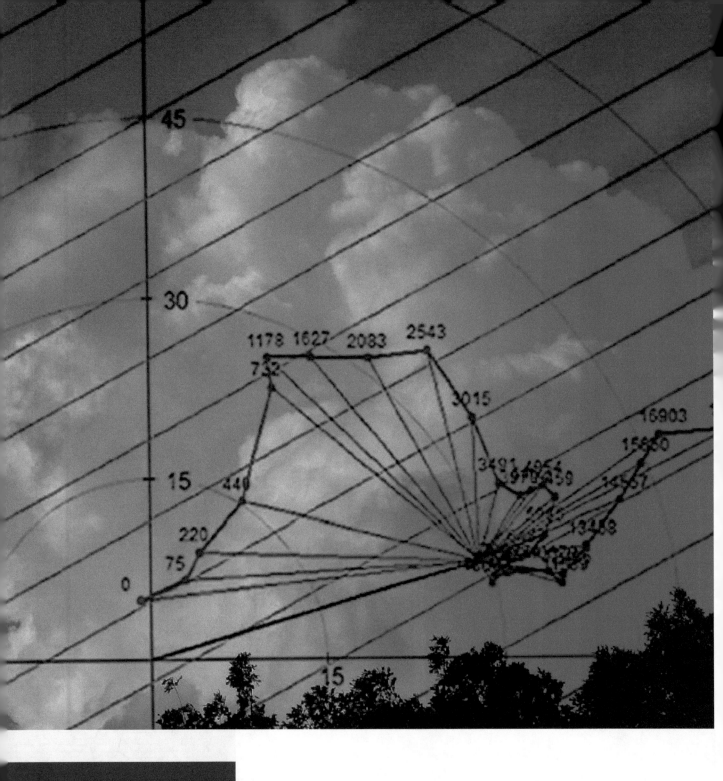

# 4 SHEAR & HODOGRAPH

The rise of mesoscale meteorology and storm modeling has shown that even thunderstorms are strongly influenced by the wind patterns in the troposphere. The change of the wind vector, in other words, the change of speed and/or direction, is a quantity known as shear.

This shear can be a horizontal measure, such as $\Delta v/\Delta x$, that is, the change in velocity per unit distance, or it can be a vertical measurement, $\Delta v/\Delta z$, the change in velocity per meter difference in height. In severe weather forecasting we are almost always dealing with the latter, the change in height. Since the change is taking place in the vertical, we refer to this as vertical shear.

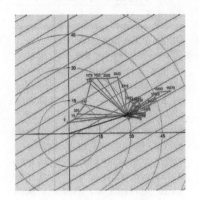

## 4.1. The hodograph

There are many different ways to plot wind data. The simplest and most familiar method to forecasters is plotting the wind barbs on the right side of the Skew-T diagram. When plotted with this method, wind barbs at the top of the diagram represent winds at high altitudes, while plots close to the bottom represent low-level winds. Shear is estimated by looking at the change of the wind speed and the turning of the wind shaft.

However this method is is difficult to use and is unreliable. For example, the profile may show substantial directional change in the winds, but only weak velocities of 5 kt throughout the column, which gives the impression of much greater shear than actually exists. It is also not possible to make calculations of shear using wind barbs.

A much more sensible technique is to plot wind information in polar coordinates. A given wind direction and velocity is plotted on the graph at an angle and distance from the graph center corresponding to the wind direction and speed, respectively. After plotting all of the wind data observed by a radiosonde, the points can be connected in "dot-to-dot" fashion.

This type of diagram is known as a *hodograph*. It provides an immensely valuable tool for visualizing changes in the wind field with altitude. For severe weather forecasters, it is a tool just as important as the Skew T digram. The hodograph is a relatively recent addition to severe weather

> The advent of tornado forecasting is undoubtedly one of the most ambitious ventures into operational weather prediction since the inception of the science of meteorology. The climatological expectancy of tornadoes and the probability of precise forecast verification is indeed so low, that for years [during the early 20th century] a "Hands Off" policy was deemed the only proper manner to treat this weather phenomenon.
>
> R. WHITING & R. BAILEY, 1957
> Eastern Air Lines

**Origin of the hodograph**

Whatever may be the complication of the accelerating forces which act on any moving body, regarded as a moving point, and, therefore, however complex may be its orbit, we may always imagine a succession of straight lines, or vectors, to be drawn from some one point, as from a common origin, in such a manner as to represent, by their directions and lengths, the varying directions and degrees (or quantities) of the velocity of the moving point: and the curve which is the locus of the ends of the straight lines so drawn may be called the hodograph of the body, or of its motion, by a combination of the two Greek words, *hodos*, a way, and *grapho*, to write or describe; because the vector of this hodograph, which may also be said to be the vector of velocity of the body, and which is always parallel to the tangent at the corresponding point of the orbit, marks out or indicates at once the direction of the momentary path or way in which the body is moving, and the rapidity with which the
body, at that moment, is moving in that path or way.

WILLIAM R. HAMILTON, 1847
*"The hodograph, or a new method of expressing in symbolical language the Newtonian law of attraction."*
Proceedings of the Royal Irish Academy

The hodograph profile is normally drawn through the depth of the troposphere. The longer this line is, the greater the shear in the atmosphere. In general, the greater the shear, the greater the potential for convective organization and long-lived storm modes.

forecasting and has proven to be a vital part of identifying supercell and tornado environments (Doswell 1991).

## 4.1.1. Rules of the hodograph

A hodograph's origin (x=0, y=0) is always fixed to the frame of measurement, and in operational forecasting this is almost always *the Earth's surface*. It is important to remember that wind data from all weather observation systems are encoded with respect to where the *wind originates from*. This is true of surface observations, radiosonde data, VAD wind profiles, and many other data sources. If an anemometer registers winds from the southeast at 15 kt, this is reported as 135° at 15 kt.

However the hodograph is a vector diagram. On a vector diagram, all arrows carry a quantity from the base of an arrow to the arrow tip (*vector* is from the Latin word "carry"). Therefore on the the hodograph, the wind quantity above is plotted at 315° at 15 kt, indicating *where the mass is moving toward*. This often confuses beginners.

Also, nearly all observed winds in meteorology are *ground-relative*. Therefore the faster the wind is relative to the ground, the further it will be plotted from the hodograph center. A calm wind is plotted at the center of the diagram.

Typically the wind profile through the entire depth of the troposphere is plotted on this hodograph. Therefore, any point on the hodograph corresponds to a specific altitude or height. By connecting a line from the diagram origin (the ground) to a specific altitude and observing the direction and magnitude of this vector, we can measure the direction and speed of the wind.

Also it is essential to remember the wind direction received in upper-air reports is reported with respect to where the wind originates from. When working with hodographs, winds are always treated in terms of which direction they are moving toward. Therefore a "southwest wind", originating from 225°, is plotted on the hodograph as 045°, in the top right of the diagram, indicating the wind is blowing from southwest toward the northeast.

Winds are also treated as vectors. Therefore when we plot a dot on the hodograph, a line between the diagram origin and the dot produces a ground-relative wind vector, indicating the velocity of the wind with respect to the ground.

An important convention in use of the hodograph is that each point of the hodograph profile should be labeled with its altitude. This is typically in km, but can be given in

SHEAR & HODOGRAPHS  105

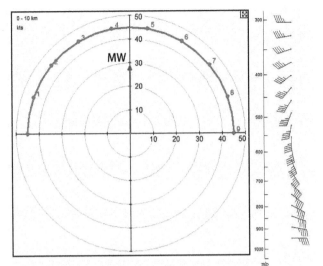

**Figure 4-1a. Semicircle hodograph**, with 180° of wind direction change through the layer and a constant ground-relative wind speed of 45 kt. Since direction is the only parameter that is changing, the wind barbs at the right show a gradual change which is easy to interpret.

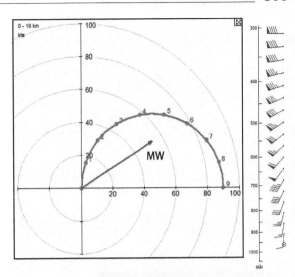

**Figure 4-1b. Semicircle hodograph**, exactly the same as the previous figure, but with an offset of 45 kt. Note that the wind barbs provide an very incomplete picture of the wind profile, showing little evidence of the strong curvature.

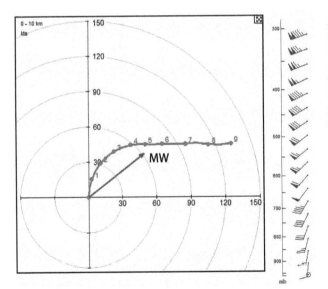

**Figure 4-1c. Curved hodograph becoming straight line with height**. This configuration is very common during supercell events, showing low-level curvature and increasing winds associated with an upper-level jet jet stream.

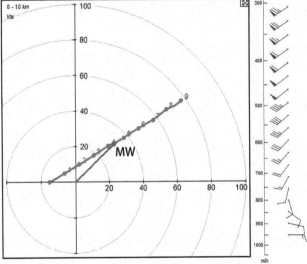

**Figure 4-1d. Straight-line hodograph**. Note how the wind barbs show significant directional change but weak mid-level winds. The mean wind vector lies directly on the wind profile, indicating very poor potential for storm-relative helicity. Splitting storms and MCSs are common with this type of hodograph, with a very low probability for supercells.

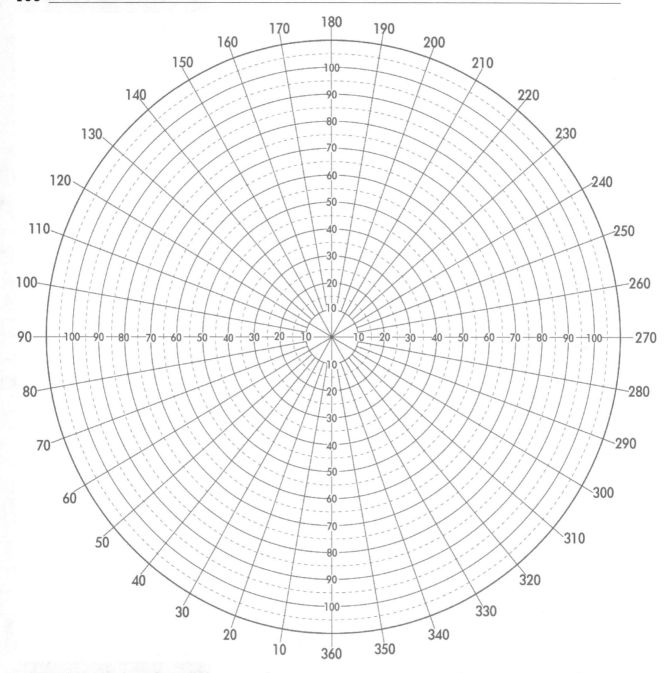

**Figure 4-2. Blank hodograph.** This diagram may be freely copied. Units may be either m s⁻¹ or kt.

feet or mb (hPa). Without this altitude information, we are severely limited in the kind of information we get from the hodograph; we only get approximate shear profiles through various layers. Omitting the height information is an oversight commonly seen on Internet graphics, so it can be helpful in many cases to generate one's own hodographs using available tools.

## 4.2. Shear

Shear is the measure of a velocity gradient. On a hodograph, which shows the wind profile above a given location on Earth, we are measuring shear along a vertical axis, so we refer to this as vertical shear. Any two points on a hodograph profile represent two different altitudes, forming a *layer* between the two points.

### 4.2.1. Bulk shear

*Bulk shear*, $\Delta V/\Delta Z$, for a layer describes the difference in velocity between the bottom and top of the layer. Using the hodograph, we select the points on the hodograph profile for the bottom and top of the layer, and plot a shear vector between those points. This gives a *shear vector*, consisting of a direction and magnitude. The *magnitude* is given as a velocity proportional to the length of the vector, and can be measured in kt or m s$^{-1}$. However it's important to emphasize that this is shear magnitude, not shear. If the magnitude is divided by the depth, $\Delta V/\Delta Z$, this gives shear through the layer in kt/m, or m s$^{-1}$ per m in which the meters cancel out to give us reciprocal seconds, s$^{-1}$. For instance, if the gradient is 50 m s$^{-1}$ over a 1000 m depth, this gives a shear of 0.05 m s$^{-1}$ per m, or 0.05 s$^{-1}$.

### 4.2.2. Total shear

*Total shear*, S, is the sum of the shear through a layer, not as a single vector between the top and bottom. It's easier to understand this graphically using the hodograph. Instead of using a single vector connecting the top and bottom of the layer, the length of the hodograph line through the layer is measured. This is equivalent to dividing the layer up into multiple layers, measuring the bulk shear magnitude in each layer, and taking the sum of all these measurements. In fact, this is the technique used by analysis programs. Typical 0–4

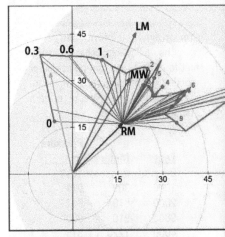

**Figure 4-3a. Hodograph for Norman, Oklahoma** on May 3, 1999 at 7 pm, at the time of tornadic storms just west of the station. Vectors have been drawn from the storm motion (RM, right motion) frame of reference. Note that this implies the storm ingests air that seems to originate directly from the west; the length of that vector indicates the wind speed (22 kt).

**Figure 4-3b. Storm-relative helicity (SRH)**; same as Figure 4-3, however SRH has been shaded in. This shows how it is constructed within an area formed by the 0 and 1 km storm-relative motion vectors. Helicity was quite large: 0-1 km SRH = 244 m$^2$ s$^{-2}$.

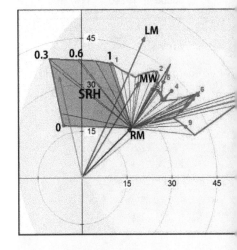

### Converting wind information

The following table converts feet to meters, and can be useful for working with hodographs. Pressure is typical observed pressure, using a standard atmosphere.

| Feet | Meters | Pressure (mb) |
|---|---|---|
| 0 | 0 | 1013 |
| 1000 | 300 | 977 |
| 2000 | 610 | 942 |
| 3000 | 910 | 908 |
| 4000 | 1220 | 875 |
| 5000 | 1540 | 843 |
| 6000 | 1830 | 811 |
| 7000 | 2130 | 782 |
| 8000 | 2440 | 753 |
| 9000 | 2740 | 724 |
| 10000 | 3048 | 697 |
| 11000 | 3350 | 670 |
| 12000 | 3660 | 644 |
| 13000 | 3960 | 619 |
| 14000 | 4270 | 595 |
| 15000 | 4570 | 572 |
| 16000 | 4880 | 549 |
| 17000 | 5180 | 527 |
| 18000 | 5490 | 506 |
| 19000 | 5790 | 485 |
| 20000 | 6100 | 466 |
| 25000 | 7620 | 376 |
| 30000 | 9140 | 301 |
| 35000 | 10670 | 238 |
| 40000 | 12200 | 188 |
| 45000 | 13700 | 147 |
| 50000 | 15240 | 116 |

km total shear in $0.05$ s$^{-1}$ is typical shear for nonsevere storms while $0.08$ s$^{-1}$ is the mean for supercell shear. Forecasters don't use these expressions of shear, however.

In operational forecasting, instead of dealing with total shear, it's often easier to measure shear as the vector difference between the wind at two different levels. By dividing the atmosphere into thin layers and observing the shear vector through each layer, we get very useful indicators of storm behavior. If all the shear vectors are pointing in the same direction, we have a "straight-line hodograph", which favors splitting modes. If the shear vectors show curvature, turning clockwise with height in the northern hemisphere, we have cyclonic shear. This favors right-flank development and rotating structures.

## 4.3. Storm motion

Storms are strongly influenced not only by the character of shear in the atmosphere, but the shear relative to the storm cell itself. Therefore, forecasters must pay special attention to the techniques for storm motion. Regardless of which method is used, a variety of different storm motions may occur in the same environment, as a result of differences such as the phase of a storm's life cycle, internal dynamics. storm depth, and other factors. The following techniques are intended to help select a most representative motion for use in storm severity algorithms and indices.

The 0-6 km vector gives an approximation of the "steering flow" in the lower and middle troposphere. Generally nonsevere storms will follow this vector. The initial storms that form during a severe weather event also take on this steering flow motion.

However forecasters deal with thunderstorms in their severe phase. When this happens, they become what we refer to as a deviant mover. Typically this is a right-mover in the northern hemisphere, but on occasion left-movers can occur, particularly in splitting modes. The forecaster's task is to anticipate and predict this deviant motion in order to calculate quantities like storm-relative helicity, which we will explore later.

Storm motion may also be affected by motion towards new updrafts (propagation), boundaries and terrain, the internal dynamics, and so forth, and again forecasters should bear in mind there may be significant differerences between various cells in the same area at the same time. This variation in cell motion is what causes cell collisions and mergers.

### 4.3.1. 30R75 method

A common method for forecasting severe storm motion is the 30R75 method, used in the 1980s and 1990s. It still can be used for a quick approximation of storm movement. It assumes that storms will move 30° to the right (NH) at 75% of the mean 0-6 km flow.

Most of the time this method works, however the shortfall of this method is that it is dependent on a ground-relative frame. As this technique is not Galilean-invariant, the method fails in "hard right movers", particularly in situations of extreme instability, or in weak flow. For example if the mean wind is 0 kt due to strong easterly flow at the surface and strong westerly mid-level winds, we find that 75% of 0 kt equals 0 kt. However a right motion vector of 0 kt is not appropriate in this situation; movement in a southerly direction at 10 to 20 kt would actually be warranted. The Rasmussen & Blanchard and Bunkers methods work properly in these conditions.

### 4.3.2. Davies & Johns method

This method (Davies and Johns 1993) is based on 31 supercells east of the Rockies. It uses the 30R75 method with a 20R85 adjustment for mean winds of under 30 kt. To explain further, the storm motion vector is calculated as 30° to the right of the 0-6 km vector at 75% of the mean wind speed, thus if mean winds are toward 060° at 40 kt, storm motion will be is 090° at 30 kt. However if this mean wind speed is 15 m s$^{-1}$ (29 kt) or less, a vector of 20° to the right of the 0-6 km vector at 85% of the wind speed is applied instead.

This provides an improved movement for storms in weak wind environments, particularly those in late spring and summer. However as with 30R75 it is not Galilean-invariant and is bound to the problems caused by using a ground-relative frame.

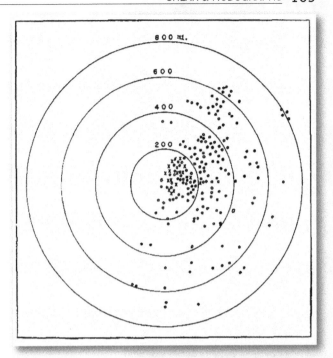

**Figure 4-4; Tornado locations** relative to center of surface low, at center, during the month of May as identified in an early study of tornadoes from 1950-1955 (Whiting and Bailey 1957). This shows the preference for the northeast quadrant, where low-level flow tends to be backed. This was not understood until the storm modeling of the 1980s. The preference of these regions for tornadoes, is now encapsulated in the parameter storm-relative helicity. Shallow mesolows with a strong response at 0-1 km but weak 2-3 km flow tend to produce strongly curved hodographs.

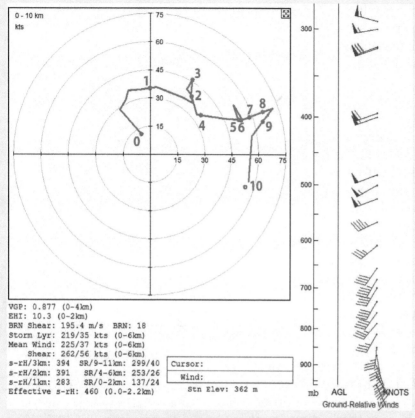

**Figure 4-5. Hodograph associated with the El Reno storm of May 31, 2013** (KOUN 01/0000 UTC), concentric range in kt. Heights are labeled from 0 to 10 km AGL; this is an important part of making a hodograph useful. The curved low-level hodograph is immediately apparent. Note that the 2, 3, and 4 km plots are closely stacked together, as is the 5 and 6 km plots. This is important because the 0-6 km mass weighted layer will be biased heavily toward the east-northeast. A pure geometric center of the 0-6 km layer is roughly at the 030/25 location, but the heavy weighting of the 2-4 and 5-6 km points places the mass-weighted average more firmly in the 045/35 range. Being able to visualize vectors in this manner is a skill that helps crosscheck diagrams for accuracy.

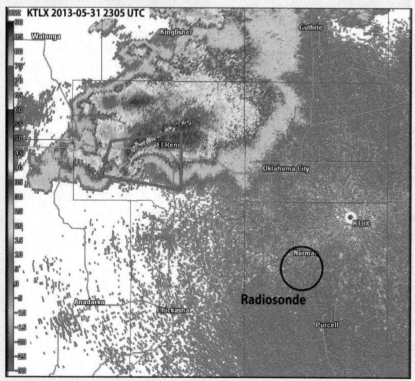

**Figure 4-6. Radiosonde launch relation to the El Reno storm.** The TTBB coded data indicated that the balloon launch occurred at 2301 UTC, and at a 5 m/s rate it would reach 1 km by 2305 UTC. Here we see a KTLX WSR-88D scan for 2305 UTC, confirming that the balloon launch is providing a fairly good sample of the inflow into the El Reno storm. This frame shows the storm shortly before the large tornado began its destructive eastward march across the region just southwest and south of El Reno.

**Figure 4-7. Construction of the ID motion (Bunkers method).** The tips of the two gray arrows shows the mass-weighted average of the wind in the 0-0.5 and the 5.5-6 km layers. A bulk shear base line (thick blue line B-B') is drawn connecting the two, which essentially provides a 0.25-5.75 km shear vector. Separately, a mass-weighted average of all points from 0 to 6 km is determined, yielding a mean wind (MW) of 040°/35 kt. A perpendicular line (orange, P-P') is drawn at a right angle to B-B' which intersects the NS point. The left mover (LM) and right mover (LM) are plotted at 14.6 kt interval along the perpendicular line. The radial scale can be used to estimate the 14.6 kt distance. Often many Internet sites will omit the LM movement as well as the NS. It is important to be aware of which vector you are working with as often these are just labeled "storm motion". Being able to work the diagram visually can help you resolve this information. By comparison, initial El Reno cell motion was given as 070°/30 kt in the severe warning, and 074-118° at 17-33 kt for the tornado warnings.

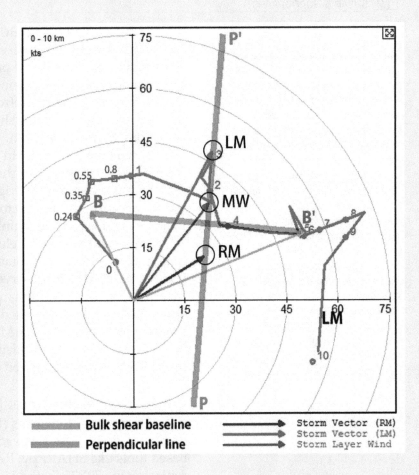

**Figure 4-8. Visualization of storm-relative helicity.** Note how the helicity is measured with triangular segments originating from the RM (right-mover) storm motion, connecting the 0 to 1 km environmental winds. The geometric area of this box (orange shaded area) is proportional to storm-relative helicity for that layer. In this case, the 0-1 km storm-relative helicity was determined to be 283 m$^2$/s$^2$. This is a large but not exceptional amount. This estimate is also only valid for the radiosonde launch site. Different wind field may exist closer to the storm. Other data, such as NEXRAD VAD, can be used to adjust the radiosonde data.

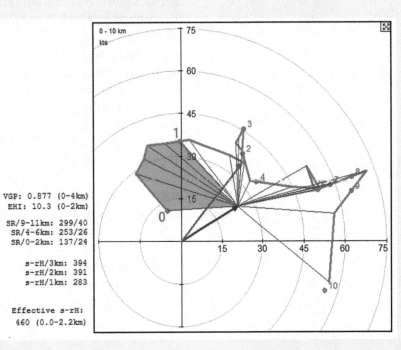

For BRN values of about 45 or less, the strongly sheared environment is crucial in producing a steady, persistent rotating updraft. This occurs as the ambient vertical wind shear and enhanced horizontal convergence increase the horizontal vorticity, which then is tilted into the vertical updraft. Due to mass continuity considerations, vertical divergence is required resulting in an accelerating updraft with height (i.e., vertical stretching). This, in turn, increases the vorticity about a vertical axis causing the development and strengthening of the mesocyclone. The strong rotation then induces a dynamic lowering of the pressure within the storm which further enhances the steady-state updraft.

### 4.3.2. Rasmussen & Blanchard method

This method (Rasmussen and Blanchard 1998, or RB98) is a simple technique based on the 0–4 km wind vector for the storm environment. Using a hodograph wind profile, a wind vector, $S_1$, is drawn from the 0 km to the 4 km wind profile. A point, which we shall call $X$, is drawn 60% of the way along $S_1$ from 0 to 4 km. A second vector, $S_2$, is drawn orthogonal (at a 90° angle) to $S_1$, starting at $X$ outward (away from the wind profile) to a distance of 8.6 m s$^{-1}$ (16.7 kt). A vector from the sounding origin (0°,0 kt), to the tip of $S_2$ yields the storm motion vector. This method is Galilean-invariant and is reliable in unusual wind conditions.

This method was developed based on an analysis of 45 hodographs which were rotated so that their 0–4 km wind vectors matched. The observed storm movement always fell within 4 m s$^{-1}$ (7.8 kt) of the motion given by the technique above. The method was used at least for several years during the late 1990s in the NCAR MM5 model, a forerunner of the WRF, and it was used to calculate severe weather parameters in the RB98 study mentioned throughout this book.

### 4.3.3. Internal dynamics (ID) (Bunkers) method

The internal dynamics (ID) or Bunkers storm motion method is widely used nowadays in online and computer based tools like SHARPpy, RAOB, and various web sites, replacing the 30R75 method. Like the Rasmussen and Blanchard method, it is a dynamically-based Galilean-invariant storm motion method, and can provide reliable estimates even with unusual wind profiles. It was developed in 1998 by M. Bunkers, B. Klimowski, J. Zeitler, R. Thompson, and M. Weisman, of the NWS, SPC, and NCAR. The technique was originally published in an AMS Severe Local Storms conference preprint, but was considered innovative enough to be formally published in 2000 in the journal *Weather & Forecasting* (Bunkers et al 2000).

The Bunkers technique solves the problems with numerous past techniques that calculated motion relative to the ground. It is related to the flexible Rasmussen and Blanchard 1998 (RB98) technique but was developed with a much larger set of supercell hodographs and uses a mean wind vector modulated by shear vectors, whereas RB98 uses a shear vector and an offset entirely. The technique is based on a study of 290 supercell wind profiles.

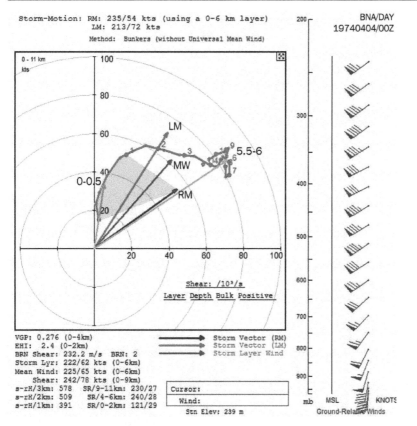

**Figure 4-9: Hodograph for April 3, 1974 superoutbreak** for east central Kentucky, based on composited Nashville-Dayton upper winds and adjusted surface wind. The hodograph shows winds rapidly increasing from the surface to 3 km, with all winds above this level between 70 and 90 kt. The very strong winds aloft result in a very high ground-relative storm motion, northeast at 65 kt for initial storms and northeast at 54 kt for right-movers. The hodograph also shows very high shear between 0 and 1 km. This high shear combined with a right-motion vector far off the hodograph results in an exceptionally high 0-1 km SRH of 391 $m^2$ $s^{-2}$ and nearly 600 $m^2$ $s^{-2}$ 0-3 km SRH.

We shall describe the technique below. Keep in mind that although it appears complicated, it is easy to work with after enough practice.

**\* Bunkers method construction.** First, the *mean wind* motion vector is estimated by taking the mean wind from 0 to 6 km, without applying weighting (Figure 4-7). This can be done visually by examining all of the sample points from 0 to 6 km, ideally using equal intervals such as every 1 km, discarding the shape and using only the points, and imagining a location at which all of these constituent points would "balance" on the tip of a pencil. This point is considered the mean wind ($V_{MW}$). Second, a mean of the 0-0.5 wind (*B*) and the 5.5-6.0 km wind (*B'*) are identified. A line is drawn between the two, forming a shear vector (*B-B'*) that acts as a base line.

Third, a line (*P-P'*) perpendicular to (*B-B'*) is drawn, which intersects $V_{MW}$. Along this perpendicular line *P-P'* the left mover and right mover ($V_{LM}$ and $V_{RM}$) are plotted. These are spaced at a distance of 7.5 m s$^{-1}$ (14.6 kt) either side of $V_{MW}$ using the diagram scale to obtain the correct scale.

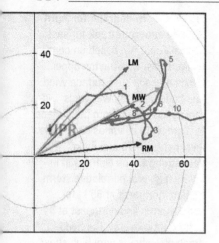

**Figure 4-10. Tornadic hodograph** for Nashville, Tennessee at noon on April 16, 1998. Note the very strong 0-1 km shear and the curvature in the 0-2 km layer. Also noteworthy is the "folding" of the hodograph back on itself with height; the 8, 9, and 10 km layers are very close to the motion vector (RM). This highlights an affinity for HP supercell modes. As the storm-relative anvil vector is very weak, the anvil wind diagnosis (UPR) in RAOB is placed very close to the 0,0 hodograph origin, indicating 17 kt of SR anvil winds placing it firmly in the HP category.

**\* Bunkers method use.** The vector from the origin to the $V_{MW}$ mean wind vector provides a base estimate of storm motion when updrafts are rooted in the boundary layer but dynamic effects and the resulting deviant motion *are not* taking place within the storm. Early during the day when storms are not surface-based, storms move similar to the $V_{MW}$ direction and speed but faster, with bias toward the mid and upper level winds.

The vector from the diagram origin to $V_{RM}$ provides "right mover" storm motion. This assumes storms are based in the low levels, and that they are being affected by dynamic pressure perturbations and are moving to the right of the mean flow. Severe storms can be expected to follow this motion, therefore storm-relative helicity and other parameters are computed using this vector. Again, the storm must be rooted in the boundary layer, otherwise the $V_{RM}$ vector will not be representative.

The vector from the diagram origin to $V_{LM}$ provides the best guess of storm motion for left-moving storms, which do in fact occur in many storm situations but are weak or short-lived. If the $V_{LM}$ SRH magnitude is higher than the $V_{RM}$ SRH, the left movers may last longer and present a greater hazard.

The problem with the Bunkers method is that due to its use of the 0-6 km layer, it is constructed for surface-based storms. The method fails with elevated storms. The effective layer Bunkers or effective layer ID method has been developed as an improvement of the original algorithm.

Also stronger mesocyclones associated with strong bulk shear are likely to produce greater deviations from the mean wind than the prescribed 15 kt. Weak mid-level flow is also more likely to produce a significant cold pool and force the storm to move more strongly with the upper-level flow.

### 4.3.4. Corrections

The forecaster normally has only the 0000 and 1200 UTC data to work with. As the atmospheric field is constantly changing, storm motion vectors have to be adjusted to be valid for the time of convection.

Wind vectors can be corrected based on radar-observed VAD/VWP winds. Surface winds give an idea of what is happening at the bottom of the column.

It is also possible to use model data, particularly from mesoscale models like the WRF and the HRRR models. Some

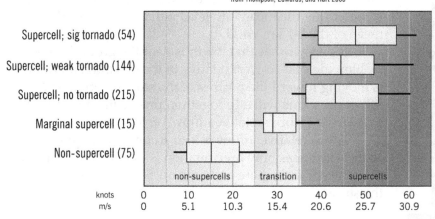

**Figure 4-11. 0-6 km bulk shear verification** of 503 events from Thompson et al 2003. This shows that bulk shear has strong skill at identifying supercell environments, but there is almost no differentiation between tornadic and nontornadic storms. The threshold values from this study indicate less than 25 kt for non-supercell storms, 25-35 for weak supercells, and above 35 kt for supercells. (Adapted from Thompson et al 2003)

sites on the Internet offer a visualization of the wind field at the forecast time.

Quite often, storm motion can be determined simply by observing convection which has already begun. It is important to differentiate elevated convection from deep convection. Elevated convection tends to form early during the day, moving quickly at approximately the speed of the 700 to 500 mb wind, and produces little severe weather. Deep convection, once it appears, has a distinctly slower movement compared to other elements on the radar, and produces intense echoes.

## 4.4. Storm-relative frame

We get a much more accurate picture of motion within a storm by considering it from a storm-relative frame of motion, as if we were somehow floating with the cell in a hot-air balloon. We can consider the entire hodograph from a storm-relative frame by simply considering the plotted storm motion to be the origin for the diagram, removing the importance of the ground.

Using this storm-relative frame, we can plot storm-relative winds simply by drawing a vector from the storm-motion origin to any level in the storm. These storm-relative wind vectors provide many indicators of the character of the storm.

**Cellular vs. linear modes**

A commonly asked question is how to anticipate whether storms will evolve into cellular or linear structures, given the existing shear and instability profiles. The strongest contributor to a linear structure is the development of a large, common outflow pool.

Weak capping with convective temperatures reached early in the day are a very strong contributor to squall lines and MCSs, as this results in the creation of numerous storms and the formation of extensive outflow. Extensive dynamic lift caused by the approach of a large trough and height falls destabilizes the column and can produce numerous cells early in the day.

Hodographs with straight-line characteristics will produce splitting storms. The appearance of both left and right movers will accelerate the development of a large cold pool.

Finally, storm movement parallel to boundaries rather than orthogonal will favor linear modes. As a result, new cells will develop along the boundary and interact with existing cells which remain close to the boundary.

**Box and whiskers**

When validating indices in this section, we use a five-number principle: the mean, the minimum and maximum (forming a range), and the upper and lower quartiles (forming a spread, or a box). The concept of the box is that it contains 50% of all the occurrences. The upper and lower whisker forms two of the four quartiles, with 25% of the highest occurrences in the upper whisker and 25% of the lowest occurrences in the lower whisker. All this makes it easy to see at a glance the distribution of the results. A small box indicates tight grouping with very similar numbers, and can indicate a good predictor, especially when it is positioned differently from other predictors.

For example, storm-relative inflow is a very important quantity that has a strong correlation with storm intensity. We draw a line from the storm motion origin point to a point representing the mean low-level wind flow, typically the mean of the 0-1 km wind.

It's easy to see that if winds near the ground are easterly, blowing east-to-west relative to the ground, and the storm is moving toward the northeast, the storm motion and the low-level winds will add together, producing a very strong east-to-west storm-relative inflow vector.

Anvil-relative inflow is also an important quantity, as this expresses how precipitation is lofted away from the storm itself, and where and how outflow will be generated at the surface.

## 4.5. Shear and storm forecasting

As early as the 1950s, shear was recognized as important for prediction of severe weather by U.S. Weather Bureau and U.S. Air Force forecasters. It isn't clear when *directional* changes were identified as a valuable parameter, but measures of directional shear were in use by the late 1960s in the earliest incarnations of the SWEAT index (Bidner 1970), the first prediction algorithm designed for tornadoes.

It was the numerical simulations of the 1980s (Weisman and Klemp 1982, 1984) which fully explored the role of shear in storm development. These types of simulations and additional studies from the operational forecasting community established parameters like storm relative helicity and the energy-helicity index, that are now in widespread use and have brought significant improvements to storm forecasts.

### 4.5.1. Deep layer shear

Bulk shear is often calculated through the 0-6 km layer, which approximates the surface through the lowest 20,000 ft and constitutes most of the storm depth. Bulk shear is computed by obtaining the vector difference between the 0 and 6 km AGL wind. Some references suggest using the mean 0-500 m winds for the bottom of the 0-6 km layer. This quantity, bulk shear, is sometimes referred to as *deep-layer shear*.

There is a strong correlation between severe weather and the magnitude of the bulk shear vector. Values of greater than 20 m s$^{-1}$ (40 kt) are generally associated with

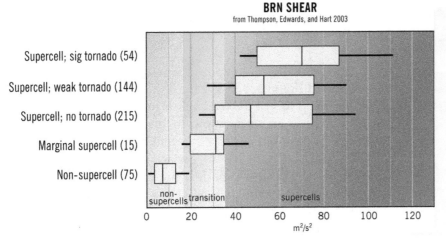

Figure 4-12. **Bulk Richardson number shear** (BRN shear) for 503 events from Thompson et al 2003. This is the denominator in the bulk Richardson number (BRN) equation, and divides the CAPE value. This yields similar results to bulk shear, with 18 and 35 $m^2$ $s^{-2}$ for the transition values. *(Adapted from Thompson et al 2003)*

severe weather. A study of 6000 soundings (Rasmussen and Blanchard 1998, or RB98) found that nonsevere storm events are associated with 0-6 km bulk shear values of 10.8 $m^2$ $s^{-2}$ (21 kt), while nontornadic supercell and tornadic supercell events showed values of 19.1 and 18.4 $m^2$ $s^{-2}$ (37 and 36 kt). Craven and Brooks 2004 (CB04) found similar results in a study of 60,000 soundings for the entire United States.

This underscores the inability of bulk shear to distinguish non-tornadic supercells from tornadic superells (Thompson et al 2002). It has been shown, however, that bulk shear may have some use in identifying *significant* tornado situations (CB04), who found median values of 24 $m^2$ $s^{-2}$ (47 kt).

Nevertheless, the bulk shear value is strongly related to storm organization, as it supports separation of the downdraft from the updraft. It also supports long-lived supercells, which are capable of producing extensive severe weather as well as long-tracked tornadoes. It is a valuable parameter for identifying severe storm environments.

New research has suggested that 0-5 km is a better layer for discriminating non-supercell from supercell storms (Houston et al 2008). On the other hand, the 0-8 km bulk shear has been found to be useful for identifying long-lived supercells (Bunkers et al 2006).

**Weak bulk shear** (below 20 kt). Weak shear environments favor nonsevere modes or disorganized multicell storms. Tornadoes are uncommon. Significant storm organization

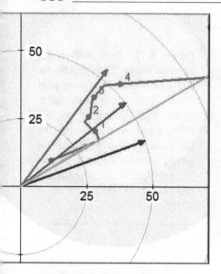

**Figure 4-13: Straight-line hodograph** for Norman, OK, April 19, 2003 at 1 pm CDT. At first glance it shows very little curvature with limited storm-relative helicity. However the magnitudes are quite large, with a 115 kt jet aloft. As a result, the sounding shows a 0-1 km SRH of 146 m$^2$ s$^{-2}$: very high. A total of 11 tornadoes touched down in Oklahoma, fortunately most were brief touchdowns.

may still occur if there are high values of shear or CAPE. The energy-helicity index (EHI) is useful for identifying these conditions in a weak shear environment. About 75% of all nonsevere storms are associated with bulk shear of 25 kt and lower.

**Moderate bulk shear** (20-39 kt). The bulk shear value of 25 kt has been found to be a useful threshold value for severe weather. This range typically encompasses organized multicell storms, but supercell potential increases through this range (Thompson 2004b), especially if conditions favor cellular instead of linear modes. Significant severe weather is associated with a mean of 35 kt or above. It has also been found that weaker shear values, 30 to 40 kt, may produce supercells in an extremely unstable environment (5000 J/kg or more) (Burgess 2003).

**High bulk shear** (40 kt or more). In high bulk shear environments, supercells are likely if cellular instead of linear modes are forecast. Linear modes will favor organized multicells, possibly bow echoes, and supercells. Significant tornadoes are associated with bulk shear of over 45 kt.

### 4.5.2. Effective bulk shear

The effective bulk shear (Thompson *et al* 2004a) provides an alternative to 0-6 km shear by using a layer bounded at the base by the most unstable layer (where theta-e is maximized), and at the top by the equilibrium level for that same parcel.

This yields values that are more tailored to each individual storm situation than the traditional 0-6 km level shear. It also reduces the bulk shear associated with elevated supercells, for which the the 0-6 km shear layer is not appropriate and tends to provide high 0-6 km bulk shear magnitudes which do not result in significant severe weather.

Note that the effective bulk shear layer is different from other effective layers such as effective CAPE.

### 4.5.3. Bulk Richardson number (BRN)

In severe weather forecasting, the bulk Richardson number (BRN) attempts to describe the balance between vertical shear and buoyancy. This is given by:

$$BRN = CAPE / 0.5(U^2)$$

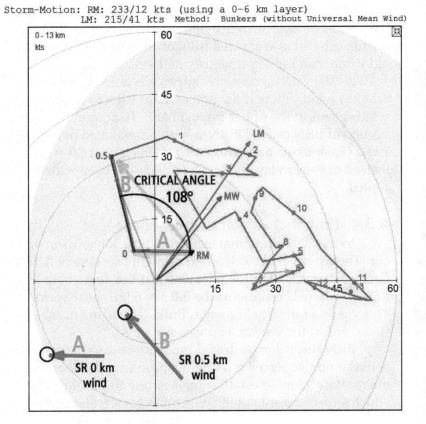

**Figure 4-14. Critical angle** is calculated from the 0 - RM vector and the 0 – 0.5 km vector. The black arc shows the angle that is being calculated. The green reference lines and the red vectors below show the storm-relative 0 km (A) and 0.5 km (B) winds. Thus, if we were riding with the storm near the surface, we would be experiencing an easterly flow at about 15 kt. Higher at 0.5 km we would experience a southeast wind at 25-30 kt. Additional information is provided in the indices chapter. Understanding this relationship is not necessary for understanding critical angle, but it is helpful.

yielding a non-dimensional number, where $U$ is the vector difference (m/s) between the density-weighted 0-6 km wind vector and the 0-500 m mean wind. As the density-weighted 0-6 km wind vector is often used as a proxy for storm motion, $U$ provides useful information about the storm-relative inflow vector.

The threshold values were originally given as 45, but we present them here as 50 in line with the findings of (Thompson et al 2003) below.

Low values, below 50, indicate strong shear in a low-buoyancy environment. Numerical simulations established that this range supports supercells.

Medium-range values, close to 50, support sustained supercells.

High values, over 50, indicate weak shear in a high-buoyancy environment. Multicell clusters or ordinary thunderstorms are likely. However field work has shown that supercells are common with BRN values of 50-100 in high-CAPE high-shear environments.

Later research using MLCAPE and 413 new case studies (Thompson et al 2003) showed BRN to be an excellent dis-

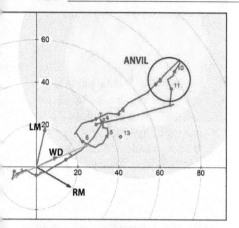

**Figure 4-15. LP supercell hodograph** for Amarillo, Texas on May 26, 1994 at 12 pm CDT. The 9-11 km winds, or anvil winds, are circled at the top right. This shows a large vector difference between the anvil winds and the storm motion vectors, on the order of 73 kt. This places the hodograph far within the LP supercell category. Indeed, picturesque LP supercells did develop.

criminator between supercells and non-supercell cases, but has almost no skill at identifying which of the supercell cases are tornadic. It also showed BRN of 50 to be a better threshold value than the 45 previously defined.

As the various classes of supercells are sensitive to 9-12 km winds, there is the potential for BRN to miss higher wind speeds in the upper troposphere. To account for this, a version of BRN called DBRN was also developed in the late 1990s (Rasmussen and Straka 1998), which uses a 0–9 km instead of 0–6 km layer. However it has not been widely accepted.

### 4.5.4. Bulk Richardson number shear

In the 2003 study that evaluated bulk Richardson number (Thompson et al 2003) an experiment was done where the CAPE value was discarded, leaving just the demominator, $0.5(U^2)$, where U represents the 0-0.5 to 6 km shear vector. This yields a quantity known as Bulk Richardson number shear, or just BRN shear, expressed in $m^2\ s^{-2}$.

BRN shear demonstrated much more discrimination between non-supercell environments and supercell environments, with 18 $m^2\ s^{-2}$ establishing a strong threshold below which supercells are not likely. Values for significant tornadoes also stand out slightly, with mean values of 70 compared to 47-53 $m^2\ s^{-2}$ for weak and non-tornadic events.

### 4.5.5. Storm-relative inflow

Storm-relative inflow appears to be related to tornado events. For surface-based storms, values below 21 kt appear to be associated with nontornadic storms, while values above 25 kt appear to be associated with significant tornadoes (Esterheld and Giuliano 2008).

Storm-relative inflow can be assessed on hodographs by observing the 0-1 km wind profile relative to the storm motion (RM) vector. The magnitude (the separation between the two on the hodograph) is proportional to storm-relative inflow. This quantity is also related to critical angle (q.v.), which measures the 0-0.5 km shear angle relative to storm-relative motion.

### 4.5.6. Inflow shear

The vertical shear *within* the inflow layer, *inflow shear*, is a good predictor for storm severity, and has been correlated with significant tornado prediction skill. Inflow shear is a measure of horizontal vorticity which can be drawn up into the updraft. In supercell environments, the curved hodograph usually shows speed shear within the layer near the ground, topped by a layer in which directional shear is dominant (Miller 2006). The transition altitude typically occurs close to the ground, around 0.5 km AGL.

In some of the earliest studies during the late 1980s and early 1990s, research focused on the 0-3 km layer. However the 0-1 km shear vector has been found to be a useful indicator of significant tornadoes; a value of 10 m s$^{-1}$ (19 kt) appears to be a useful threshold between significant tornadoes and all other thunderstorm events (Craven and Brooks 2004). This measure can be used as a proxy for storm-relative helicity when it is unavailable (Thompson et al 2002), and offers similar capabilities without the need for computing a storm motion vector.

Some newer studies have suggested that even shallower layers such as 0-0.5 km shear may provide strong discrimination between nontornadic and weak/significant tornadic events (Esterheld and Giuliano 2008). A shear value of 15 kt is suggestd to be the best threshold value. The 0-0.5 km vector can be quite strong in tornadic environments; for example, in Esterheld's paper an analysis of Dodge City WSR-88D VAD/VWP data showed that the Greensburg, Kansas tornado was associated with a shear vector of 39 kt.

### 4.5.7. Critical angle

The critical angle equals the angle between the surface storm-relative wind and the 0 – 0.5 km shear vector. In effect this examines the 0 to 0.5 km environmental winds, which comprise the storm's inflow, using the storm-relative frame of reference, usually the right mover. These winds are examined as if we are moving with the severe storm.

It has been found that critical angles over 90° are associated with weak to significant tornadoes, while angles near 120° are associated with nontornadic events (Esterheld and Giuliano 2008).

### 4.5.8. Storm-relative anvil winds

Modeling has shown that the typical level of nondivergence in severe storms is at about 9 to 10 km, and storm-relative flow above this level is especially important for "ventilating" the updraft. Likwise, this flow has been found to be strongly correlated with supercell mode. Storm-relative anvil flow expresses the magnitude difference between the storm

---

**Reading the UCAR anvil winds**

Only a handful of websites plot the storm-relative anvil winds. Among them is the UCAR Research Applications Laboratory realtime sounding plotter at:
   weather.rap.ucar.edu/upper

On the mini-hodograph at the top, The blue dot indicates the estimated supercell storm motion computed by taking 60% of the magnitude and 8 m/s (16 kt) to the right of the boundary-layer to 4 km shear vector. This blue circle represents a 4 m/s error estimate and is also indicated by the heading "CELL" along the top and is only valid for developing to mature supercell thunderstorms, not necessarily all general thunderstorms.

It must be emphasized that this is a ground-relative wind calculation, not an internal dynamics (Bunkers) method, so the storm motion is subject to error. These types of calculations are most likely to fail in slow-moving high CAPE situations.

When the blue dot lies in the white area, no supercells are anticipated. When the blue dot lies on the green ring, then high-precipitation (HP) supercells are expected. When the blue dot lies on the gray ring, then classic (CL) supercells are expected and when the blue dot lies outside the gray ring, then low-precipitation (LP) supercells are expected.

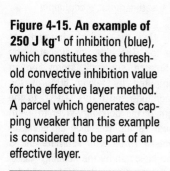

Figure 4-15. An example of 250 J kg⁻¹ of inhibition (blue), which constitutes the threshold convective inhibition value for the effective layer method. A parcel which generates capping weaker than this example is considered to be part of an effective layer.

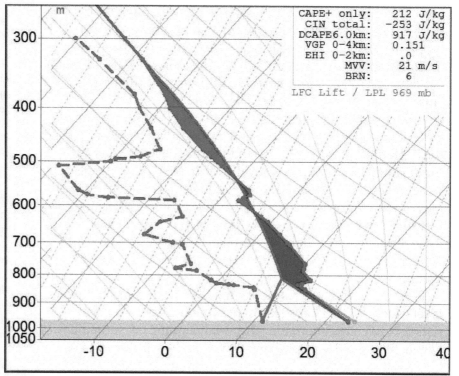

motion vector, usually the right mover RM vector, and an average of the 9-11 or 9-12 km winds (Rasmussen and Straka 1998).

**SR anvil winds less than 18 m s⁻¹ (35 kt)** are associated with high-precipitation (HP) supercells. This suggests the upper portions of the cumulonimbus cloud are generating precipitation at close distances from the updraft core, and extensive updraft-downdraft interaction will result.

**SR anvil winds of more than 28 m s⁻¹ (55 kt)** are associated with low-precipitation (LP) supercells. The precipitation is being lofted downshear, often 5 to 10 miles away, mixing extensively with environmental air and dissipating the precipitation column further. There is very little interaction between the updraft and downdraft.

**SR anvil winds of 18-28 m s⁻¹ (35-55 kt)** are associated with a range of supercell modes, from LP at low storm-relative helicity (SRH) levels, to classic, CL, at medium SRH, and to high-precipitation, HP, at high SRH. The study specified 250 m2 s-2 as a good threshold value between LP and CL, but did not specify a threshold between CL and HP.

In a classic supercell, there is a balanced relationship between the updraft and downdraft. Interaction between the two varies, primarily depending on the supercell's stage in its life cycle. The supercell usually goes through different stages during its life cycle.

**Depiction.** Storm-relative anvil winds are often portrayed on hodographs with a shaded donut-shaped circle. This may be centered on either the anvil winds (NCAR website) or on the motion vector (RAOB program). By observing where the other part of the SR anvil wind component is located, the magnitude can be quickly assessed.

### 4.5.9. Corfidi vectors

The Corfidi vector represents a tool for forecasting squall lines and other linear convective systems. It quantifies storm movement based on factors like the low-level jet and mean cloud winds. The initial version (Corfidi et al 1996) examined 103 mesoscale convective systems. It found that movement equaled the sum of mean cloud-layer wind (MCLW), determined by an average of the 850, 700, 500, and 300 winds, and the opposite of the low level jet. This provided an advective and propagation term, respectively, which when added together yielded storm motion.

Over the next few years, further experience found that the Corfidi vectors did not work well in rapidly propagating MCSs, prompting a revision (Corfidi 2003) that examined 48 convective systems with damaging surface winds. This differentiated *upwind-propagating* MCSs, which were forecast by the original version, from *downwind-propagating* MCSs. Refer to Figure 4-16.

* **Upwind-propagating MCSs** - Motion, $V_{MBE}$, is the sum of MCLW, $V_{CL}$, added to the opposite of the low-level jet (propagation, $V_{PROP}$, thus, advection and propagation tend to be subtractive. This motion is most likely in moist environments, with weak downdrafts. This is typical of mesoscale convective complexes, MCCs, and summer systems that bring flash flood potential. Propagation favors the side of the cold pool parallel with the mean flow.

* **Downwind-propagating MCSs** - Motion, VMBE, is the sum of the upwind-propagating $V_{MBE}$ vector added to the $V_{CL}$ vector, thus advection and propagation are additive. This mode is most likely with dry mid-tropospheric conditions

---

**MCS types**
A study by Moore et al 1993 found the following MCS characteristics:

**Upwind (backward) propagation**
* Maximum CAPE is S-SW of MCS
* East-west frontal boundary
* 850 mb theta-e axis is to the west
* Difluent thickness pattern
* Weak 850-300 mb wind (<30 kt)
* 300 mb jet is weak and well to N
* Warm advection or lack of PVA
* LLJ is normal and upstream from the surface boundary

**Downwind (forward) propagation**
* Strong 850-300 mb (>30 kt)
* Maximum CAPE is near or to east
* 300 mb jet is zonal and to north
* LLJ is coincident with the MCS

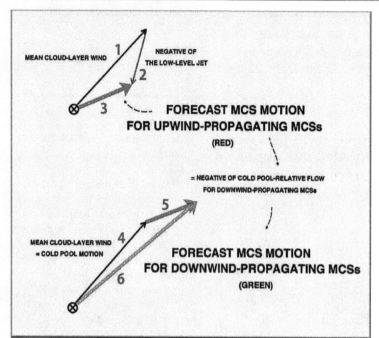

**Figure 4-16. Corfidi vectors.** The top example shows upwind (backward) propagation, starting with mean cloud-layer wind (1), opposite of the low-level jet (2), and the resulting vector (3) for upwind propagation. The down-wind (forward) propagation adds mean cloud-layer wind (4), to the upwind-propagating vector (5), to obtain the resulting vector (6) for downwind propagation. (Adapted from Corfidi 2006)

with strong downdrafts, particularly in bow echo and derecho situation. Propagation favors the side of the cold pool parpendicular to the mean flow.

## 4.6. Helicity

During the 1980s it was well-established that supercell thunderstorms developed in high instability and high shear, but the existing indicators of shear failed to differentiate nontornadic from tornadic storms. Old indices like the SWEAT index showed some skill due to its use of directional shear, but the parameter used a 850-500 mb quantity, which often overshoots the critical inflow layer.

By the mid-1980s storm modeling provided an important breakthrough: curvature in the hodographs, particularly in the lowest levels, solved the elusive distinction between tornado and non-tornado cases (Weisman and Klemp 1982-1986; Patrick and Keck 1987). The best quantification of this curvature is storm-relative helicity (Davies-Jones et al 1990), and this remains in use today.

In the late 1980s, the 0 to 3 km layer was frequently used in storm-relative helicity calculations. The 0-3 and 0-1 km layers are called fixed-layers. They provide misleading estimates of storm-relative helicity in certain situation, particularly elevated moisture as might be found poleward of frontal boundaries. Also mixing air through these layers in order to find MLCAPE can result in misleading values, so caution must be exercised.

### 4.6.1. Storm-relative helicity

Storm-relative helicity is a measure of the curvature of the shear vector. This is measured through a specific layer, and is *evaluated relative to the storm motion origin*. The geometric area swept out on the hodograph (Figure 4-8) is proportional to the storm-relative helicity. *Storm-relative helicity measures streamwise vorticity.*, which produces storm rotation when environmental low-level shear is tilted into the updraft.

The magnitude of SRH increases as storm motion moves "off" the hodograph; that is, when the storm motion differs significantly from all of the winds comprising the hodograph profile within the SRH layer, SRH magnitude will be larger. This is one reason why deviant movers are more effectively able to develop rotating updrafts and tornadoes.

Note that storm-relative helicity is different than helicity, which. Meteorologists must be careful to use the proper terminology, using SRH or "storm relative helicity" in storm forecasting and not simply "helicity".

### 4.6.2. Storm-relative helicity (0-3 km)

The earliest storm-relative helicity investigations (Davies-Jones et al 1990) used a layer of 0-3 km, based on storm modeling work in the 1980s, much of it by Weisman and Klemp. A major shortfall of the 0–3 km layer is that it ranges upward to almost 10,000 ft AGL, thus it frequently samples air above the cap which will not be ingested into the updraft. In the vast majority of Great Plains cases, 3 km lies within the southwesterly flow embedded within the capping inversion.

This deep layer, however, was sufficient to reveal the association between storm-relative helicity and severe weather. Values of 0-3 km SRH of 55, 124, 180 $m^2 s^{-2}$ are *median* values associated with nonsevere, nontornadic supercell, and tornadic supercells, respectively, in a study of over 6000 soundings (Rasmussen and Blanchard 1998). The 168 $m^2 s^{-2}$ value has been given as a transition between nonsevere and severe storms. An SPC severe weather parameter page thought to be written by R. Thompson lists 250 $m^2 s^{-2}$ as a threshold value for tornadic supercells. Elsewhere in the literature, a study by McCaul 1988 found a mean storm-relative helicity of 150 $m^2 s^{-2}$ for hurricane tornadoes, and this was cited in the Davies-Jones 1990 paper.

This original paper by Davies-Jones 1990 gave a set of threshold values for 0-3 km SRH. Values below 150 are not associated with rotating storms. Values of 150, 300, and 450 m2 s-2 indicate lower limits for weak, strong, and violent tornadoes.

Though the 0–3 km SRH parameter does provide a useful indicator for rotating storms, it is not ideal for *tornadic* storms. The 0-1 km SRH and effective layer SRH tend to be better predictors.

**Backed and veered winds**
Backing describes a counter-clockwise* tendency in the wind field, whereas veering describes a clockwise* tendency. It's important to understand the context so as not to be confused.

**Layer** - The most common use of backing and veering describes a wind direction change within a layer, from bottom to top. Providing friction or localized effects are not dominating the wind, backing in a layer from bottom to top is associated with cold advection, while veering indicates warm advection. Near the ground up to an altitude of hundreds or thousands of feet, frictional effects dominate, so veering is normally observed.

**Time** - Veering and backing can occur with respect to time. For example, if a south wind becomes southeast an hour later, this wind is backing with time. Pressure falls, indicating upper-level lift, often back the winds with time.

**Air mass** - Finally we can describe changes with respect to an air mass. For example, let's consider an east-west outflow boundary. When crossing from warm into cool air, we might find the wind changes from southeast to northeast. In this case we can describe the winds as being locally backed in the cold air. Using the "layer" method, a station in the cold air observes veering with height from northeasterly to southerly at 1 km, where a low level jet might be found. However it also observes that the surface winds are locally backed compared to what would be seen without the cold air. This is a key point for understanding storm-relative helicity as might be observed northeast of a mesolow.

* Opposite in the southern hemisphere

> **The cause of failure**
> The author's most regrettable severe storm forecast mistakes have arisen from ignoring data that were relevant to the daily diagnosis...and/or failing to complete the diagnosis on what initially appeared to be a benign weather day.
>
> ALAN MOLLER, 2001
> *Severe Convective Storms*

> **The five basic ingredients**
> Listed here are the five basic ingredients for severe storms:
> * Buoyancy
> * Deep-layer shear
> * Storm-relative helicity
> * Moisture; high RH%; low LCL
> * Discrete development

### 4.6.3. Storm-relative helicity (0-1 km)

New research in the late 1980s and early 1990s identified the 0 to 1 km layer as critical for tornadogenesis. The 0-1 km storm-relative helicity (SRH) is now used as the standard for determining SRH as it applies to tornadogenesis. Some newer research has even suggested using the 0 to 0.5 km layer to identify tornadic environments, however this technique is not in widespread use.

For 0-1 km SRH, values of 15, 33, and 89 $m^2 s^{-2}$ are associated with nonsevere, nontornadic supercell, and tornadic supercells. Other sources (SPC 2017) give 100 $m^2 s^{-2}$ for an increased threat of tornadoes with supercells. All of these values are to be used as guides only, not as clearly delineated thresholds.

### 4.6.3. Storm-relative helicity (ID / effective layer)

The effective layer ID method (Thompson et al 2004a) uses the effective layer instead of a fixed layer, as with the original Bunkers method. This method only uses layers with positive buoyancy. In common usage, there is a pairing of conditions: for a layer to be considered buoyant, CAPE must be greater than 100 $J kg^{-1}$ and CINH must be weaker than 250 $J kg^{-1}$ (i.e. lower than this number). This CINH corresponds to a fairly strong cap.

The algorithm starts at the surface and ascends, finding the first level meeting the buoyancy criteria. This defines the effective layer base. In a warm, moist environment, this level frequently equals the surface. The algorithm keeps ascending, and if an effective layer base has already been found, the first layer which fails to meet the buoyancy conditions is considered the effective layer top.

The effective layer can be constrained by increasing the CAPE threshold. By raising it from 100 to 500 J/kg, the effective depth is often reduced by several hundred meters (Thompson 2007).

It is important to remember that the majority of soundings, especially in stable cool weather conditions, will fail to find any effective layer. If this is the case, effective layer computations cannot be made.

Care should be used when abbreviating "effective layer" to EL, as this can be confused with "equilibrium level".

# Review Questions

1. Does the direction of the vector corresponds to where the air has come from (METAR and SYNOP reported direction) or where the air is going toward?

2. What does the length of the hodograph line correspond to?

3. What does curvature of the hodograph line correspond to?

4. What is the difference between bulk shear and total shear?

5. How is 0-6 km bulk shear calculated?

6. The Bunkers method plots the right and left mover how far from the base line?

7. What is the instability criteria for the effective layer when working with effective layer SRH?

8. Early in the day when deep convection is forming, how do these cells move in relation to the hodograph wind vectors?

9. Describe the 30R75 movement and what its advantage is.

10. What are some sources of wind field corrections as the day progresses?

11. The 0 - 6 km bulk shear vector connects which points on the hodograph?

12. What kind of storms are associated with low bulk shear (below 20kt)

**Discrete modes**
Elements for discrete modes:
* Environment favoring isolated cell coverage (see above)
* Weak cold pool
* Boundaries with weak temperature contrasts
* Shear vectors perpendicular to boundaries

**Linear modes**
Elements which favor linear modes:
* Environment favoring extensive cell coverage (see above)
* Creation of a large cold pool
* Boundaries with strong temperature gradients
* Shear vectors parallel with boundaries

13. What kind of storms are associated with high bulk shear (40 kt or more).

14. Which layer is most important in tornado prediction?

15. Describe the critical angle.

16. Which severe weather forecasting quantity provides a measure of streamwise vorticity, which is associated with rotation?

17. Why does 0-3 km SRH not provide a useful measurement?

18. What type of storm environment is provided by a straight line hodograph?

19. Which magnitudes of storm-relative inflow are associated with severe weather?

20. Describe the Bulk Richardson Number (BRN).

# A-1 SEVERE WEATHER INDICES

Since the early years of storm prediction, forecasters and researchers have attempted to convert the raw information on the diagrams into parameters which yield immediate forecast information. A *stability index* or *forecast index* attempts to quantify the ability of the atmosphere to produce thunderstorms or other types of weather. There are also more complex *algorithms* which, likewise, predict weather but change the result according to whether specific conditions are met.

A basic index consists of either a scale or threshold values. With a scale, the likelihood of the weather event or specific phenomena increases as the index value decreases or increases. With threshold values, there is a predefined value at which thunderstorms or weather phenomena are likely to occur.

Indices are generally computed directly from the data on the sounding or hodograph. More complex index techniques use an algorithm or flowchart.

It must be remembered that indices are essentially statistical forecasting tools. They do not model dynamic processes occurring in the storm, and only provide approximations of updraft strength, updraft-downdraft separation, and so forth.

Also forecasters must be aware that complex indices are prone to failure, because each term represents an assumption or a simplified part of the thunderstorm process, and when one term fails, it can have unpredictable effects on the rest of the equation. This is why elaborate indices such as SWEAT are no longer in use. The best forecasts are created by the forecaster using simple indices, checking for consistency, and assessing the contribution of each one to the forecast.

## A1.2. Case study values

Provided with each parameter is a listing of values for that parameter during various events. It must be emphasized that no exceptional efforts were made to identify environments and parcels that were most representative for that storm environment. Therefore the values seen here are typical of what will be shown on sounding analysis tools when analyzing a forecast situation.

For most cases, the nearest sounding that represents the inflow air mass was chosen. The 000 UTC sounding was

**Objective forecasts**
Once an objective scheme is built, it become a "black box" about which one may choose to know nothing, save the output. Interestingly, when one is ignorant of the contents of such a system, it is hard to anticipate when and where it might fail.

CHARLES DOSWELL, 1986
The Human Element in
Weather Forecasting

**Tornado panic**

A forecast of tornadoes, unless properly circumscribed, would lead to undue public alarm and might seriously interrupt business and industrial activity when the localities touched by the tornadoes might be limited to isolated rural regions with no material damage. In the Weather Bureau's long history, there have been cases in which the damage caused by rumors of tornadoes was greater that caused by the storm itself.

Under the present conditions of war and greatly increased public interest in the subject, it is all the more necessary to offset such rumors by dissemination of authentic meteorological advice and to do whatever is possible in issuing specific forecasts and warnings when destructive local storms endanger life and property, especially in war concentration ... In some cases local storm reporting networks can give a few minutes warning before the arrival of severe storms when their forecasting by regular synoptic methods has not been practicable.

Experimental warning networks for this purpose have been organized in various places and will be extended as soon as practicable if the results so justify. These networks cannot be expected to provide adequate warnings in every case; nevertheless they offer the possibility of providing more specific warnings beyond what can be done by regular synoptic forecasting, and it is the Weather Bureau's responsibilities to develop these possibilities to the fullest extent.

- F. W. REICHELDERFER
Chief of the U.S. Weather Bureau
Circular Letter 55-43, June 1, 1943

used, which incorporates surface heating and environmental changes during the day. RAOB 6.8 was used to measure all the quantities.

In some locations where there is poor coverage (central Texas) two soundings were merged, and in one case the upstream low-level moisture field was "advected" to the sounding site. Where radiosondes were not available or did not adequately sample the storm air mass, the case study was rejected.

**Topeka KS F5 (6/8/1966)**. A large tornado passed through Topeka, Kansas, killing 16 and injuring 450. The Topeka 09/0000 UTC sounding appeared to adequately sample the air mass and was used.

**Superoutbreak (KY) (4/3/1974)**. An outbreak of 148 tornadoes hit the east central United States, the most numerous tornado outbreak in history. This calculation focuses on Kentucky, based on a composite 0000 UTC Nashville-Dayton sounding with surface data.

**DFW Texas / Delta 191 (8/2/1985)**. On the evening of August 2, 1985, Delta Flight 191 crashed on approach to Dallas-Fort Worth International Airport. This was caused by weakly-organized summer storms which had brief periods of intense downdraft activity, occasionally identified as *pulse thunderstorms*. A merged Stephenville and Longview sounding was used.

**Jarrell TX F5 (5/27/1997)**. An F5 tornado developed along a stagnant front in a high-CAPE low-shear air mass and moved southward. At least 27 were killed. Storm chaser footage showed unusual detail of violent, contorting subvortices. Due to the poor radiosonde coverage in central Texas, a combined Fort Worth - Corpus Christi sounding was used.

**Cen. Oklahoma (5/3/1999)**. A major tornado outbreak occurred in Central Oklahoma, One supercell tracked up I-44 from Bridge Creek to Moore, killing 36 and producing $1.4 billion in damage (2017 adjusted). The Norman 04/0000 UTC sounding was used.

**Fort Worth F3 (3/28/2000)**. An F3 tornado moved through downtown Fort Worth, Texas, producing $600 million (2016 adjusted) in damage, but fortunately with no fatalities. The

Fort Worth radiosonde, launched just east of the storm and providing a near-storm sample, was used.

**St Louis hailstorm (4/10/2001).** On April 10, 2001 an HP supercell tracked east along I-70 toward St. Louis, damaging 125,000 homes and 65,000 cars. Poor radiosonde coverage limited the ability to sample this air mass; a merged Springfield, Missouri and Lincoln, Illinois sounding had to be used.

**Greensburg KS EF5 (5/4/2007).** On May 4, 2007, a wedge tornado measuring up to 1.7 miles wide passed through Greensburg, Kansas, destroying the town and killing 11. The Dodge City 05/0000 UTC sounding was used, modified with the moisture profile from the Norman sounding.

**Joplin MO EF5 (5/22/2011).** On May 22, 2011 a violent tornado moved through Joplin, Missouri, killing 158, the highest death toll in a single storm since the 1947 Woodward tornado. A 23/0000 UTC radiosonde from Springfield provided a good near-storm proximity sounding.

**Ohio derecho (6/29/2012).** On June 29, 2012, a long-tracked derecho moved from just east of Chicago to Virginia overnight. A special 29/1800 UTC Wilmington, Ohio radiosonde launch provided the only clear sample of the pre-derecho environment.

**Moore OK F5 (5/20/2013).** Moore, a city south of Oklahoma City, has had unusually bad luck with damaging tornadoes. Once again it fell prey to an EF5 tornado which killed 24. Reinforcing the violence of this tornado is the fact that there was no dispute of the damage rating. The 21/0000 UTC Norman sounding was used.

**El Reno OK EF3 (5/31/2013).** A wedge tornado developed southwest of El Reno and was captured by numerous chasers. Eight were killed, among them tornado chaser Tim Samaras, his son Paul Samaras, and fellow chaser Carl Young. The Norman 01/0000 UTC sounding was used.

**Vilonia AR EF4 (4/27/2014).** A large wedge tornado developed south of Conway, Arkansas (about 30 miles northwest of Little Rock), with 16 fatalities. Structural engineer Tim Marshall suggested EF5 damage was possible. The Little Rock 28/0000 UTC sounding was used.

**Rochelle IL EF4 (4/9/2015).** A supercell produced six tornadoes in north central Illinois, killing 2 and injuring 22. This was the first EF4-level tornado to strike the Chicago warning area since the Plainfield, Illinois event on August 28, 1990. It appears to be officially listed at EF4 though signs of EF5 were observed. The 0000 UTC Lincoln IL sounding was used.

# CAPE | Convective available potential energy

**Purpose**
- Maximum updraft speed
- General purpose storm severity

**Abbreviations**
CAPE, PBE, +BE

**Introduced**
Mitchell Moncrieff & Martin Miller, 1976
Concept of positive area: Alvord, 1929

**Maximized**
High low-level moisture and deep, cold upper-level layers

**Units**
$J\ kg^{-1}$

**Parcel types**
* Mixed layer (ML) - All warm-sector storms; shallow moisture situations
* Most unstable (MU) - Quick identification of risk areas; all elevated storms
* Surface based (SB) - Summer storms

**Representative values**

| | |
|---|---|
| 0-500 | Weak — fair weather; or isolated weak showers |
| 500-1000 | Low — scattered showers |
| 1000-2500 | Moderate — thunderstorms likely |
| 2500-4000 | Strong — Widespread severe storms |
| >4000 | Extreme — large hail |

**Examples**

| | |
|---|---|
| 490 | St Louis hailstorm (4/10/2001) |
| 792 | DFW TX / Delta 191 (8/2/1985) |
| 869 | Rochelle IL EF4 (4/9/2015) |
| 1134 | Superoutbreak (KY) (4/3/1974) |
| 1388 | Topeka KS F5 (6/8/1966) |
| 1692 | Fort Worth F3 (3/28/2000) |
| 2092 | Vilonia AR EF4 (4/27/2014) |
| 2206 | Cen. Oklahoma (5/3/1999) |
| 2311 | Jarrell TX F5 (5/27/1997) |
| 2484 | Moore OK EF5 (5/20/2013) |
| 2795 | El Reno OK EF3 (2013-05-31) |
| 2850 | Ohio derecho (6/29/2012) |
| 3830 | Joplin MO EF5 (5/22/2011) |
| 4072 | Greensburg KS EF5 (5/4/2007) |

Out of the multitude of available indices, convective available potential energy, CAPE (Moncrieff and Miller 1976) is the most useful and versatile parameter by far, because it bears a direct relationship to the energy available to a parcel and is strong linked to severe weather. CAPE is also known as positive buoyant energy (PBE, or +BE). It does not measure instability of a layer, but arrives at updraft energy by examining temperature difference between a rising parcel and the environment and integrating these measurements with height.

As far back as the 1940s, the importance of positive area on a sounding for thunderstorm development has been well understood. Although it is not clear why a forecast parameter based on positive area wasn't developed, the lack of computation power for performing integrations was probably a reason. Forecasters instead developed fixed-level parcel temperature difference indices such as the Showalter Index and Lifted Index.

Mitchell Moncrieff and Martin Miller at London's Imperial College developed the idea of CAPE in the mid-1970s in an effort to better quantify buoyancy for a study of tropical thunderstorms (Moncrieff and Miller 1976). Its use in severe storm modeling over the next decade led to widespread replacement in operational forecasting of lifted index with CAPE in the 1990s.

The representative CAPE depends on a proper selection of the parcel and knowledge of the state of the atmosphere at initiation time. Also in certain situations, CAPE should be constructed from elevated layers. Internet Skew-Ts and even SHARPpy may not adequately capture such elevated layers, and forecasters may need to use sounding manipulation tools such as RAOB to properly evaluate them.

Values of over 8000-10,000 $J\ kg^{-1}$ are rare and near the highest ever observed; these have occurred in the central US, eastern India, and Bangladesh.

# SEVERE WEATHER INDICES 135

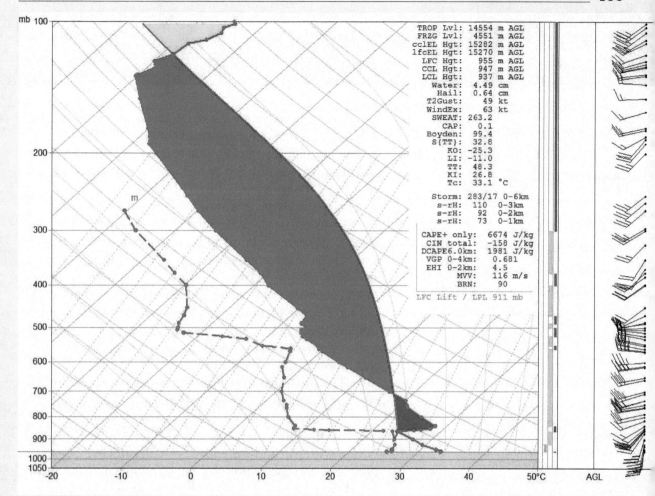

**Figure A1-1: Extreme MLCAPE**, the highest ever observed at Oklahoma City / Norman, Oklahoma. The mixed layer CAPE was 6674 J/kg, observed on the evening of July 1, 1992. This yields a rather extreme lifted index of -11. The massive red positive area depicts the CAPE. This was also associated with very high NCAPE, 0.47. The strong cap (CINH -158).

**Figure A1-2: Most unstable CAPE** as depicted horizontally on SPC mesoanalysis graphics during a cool-season marginal thunderstorm event along the Gulf Coast. (NOAA)

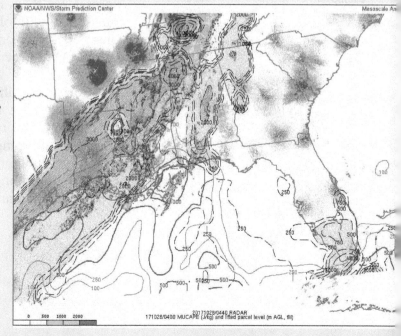

# CIN | Convective inhibition

**Purpose**
- Maximum updraft speed
- General purpose storm severity

**Abbreviations**
CIN, CINH

**Introduced**
Frank Colby Jr., 1984
Concept of negative area: Alvord, 1929

**Maximized**
High low-level moisture and deep, cold upper-level layers

**Units**
J kg$^{-1}$

**Parcel types**

**Representative values**
Absolute value, J kg$^{-1}$
<0          Uncapped
0 – 24      Weak capping
25 – 49     Moderate capping
50 - 100    Strong capping
>100        Usually unbreakable capping

**Examples**
These should not necessarily be considered representative of the near-storm environment. See chapter notes.
4       Joplin MO EF5 (5/22/2011)
8       Greensburg KS EF5 (5/4/2007)
31      Vilonia AR EF4 (4/27/2014)
38      DFW TX / Delta 191 (8/2/1985)
49      Rochelle IL EF4 (4/9/2015)
68      Fort Worth F3 (3/28/2000)
79      Moore OK EF5 (5/20/2013)
81      Topeka KS F5 (6/8/1966)
96      Superoutbreak (KY) (4/3/1974)
103     St Louis hailstorm (4/10/2001)
121     Cen. Oklahoma (5/3/1999)
125     El Reno OK EF3 (5/31/2013)
172     Jarrell TX F5 (5/27/1997)
181     Ohio derecho (6/29/2012)

Convective inhibition (CIN, or CINH) is measured in J kg$^{-1}$. It represents how much energy needs to be added to a rising parcel so that it encounters no layers that are warmer than the parcel, and is calculated in the very same manner as CAPE. Therefore the negative area on a sounding is proportional to convective inhibition.

Convective inhibition is a negative energy expression, so positive amounts of convective inhibition have a suppressive effect. Increasing the magnitude of cap increases the convective inhibition, so a station observing 75 J kg$^{-1}$ is much more strongly capped than a station with 20 J kg$^{-1}$. Some caution is required because there are occasionally charts and references that give CIN as a positive energy expression, thus making CIN a negative number, and this can cause confusion.

The earliest years of severe weather forecasting highlighted the role of a mid-level warm layer in augmenting severe weather. This came to be known as the "cap" or the "lid" (Fawbush and Miller 1952 and Beebe 1958). With the widespread acceptance of CAPE in the 1980s and the use of positive buoyant area on the sounding, it was immediately clear that negatively buoyant areas could likewise be identified and measured using the same technique used to calculate CAPE. This new expression, CIN, was first used in an 1984 paper on by Frank P. Colby Jr. of the Massachusetts Institute of Technology analyzing data collected during the multi-agency 1979 SESAME experiment, an early predecessor of VORTEX.

Correct modification of all parts of the sounding and choosing the correct parcel is especially critical in the case of convective inhibition. It cannot be used with an unmodified morning sounding. Horizontal plots of convective inhibition produced by model output, however, are useful for finding the distribution of areas that are capped and other areas that are not.

SEVERE WEATHER INDICES  137

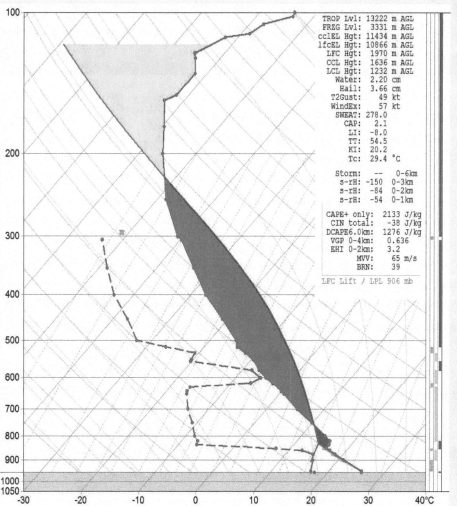

**Figure A1-3 (left): Convective inhibition** (blue area) using the 1200 UTC sounding modified with the forecast maximum temperature at the dry adiabatic lapse rate. This left a CIN of 38 J kg$^{-1}$ which was at least partially sufficient to suppress all convection. This example is for Oklahoma City on May 8, 1988, a well-known storm chaser bust day at the time.

**Figure A1-4 (below); Convective inhibition plot** (cyan), with dewpoint (green), CAPE (red) and wind field (gray). It shows an notable absence of convective inhibition, although the air mass was definitely capped at the time. The source of the problem can be seen (right): coarse vertical resolution in the NCEP North American Regional Reanalysis, where layers are every 25 mb. A resolution of 10 mb or better is necessary to properly sample the cap.

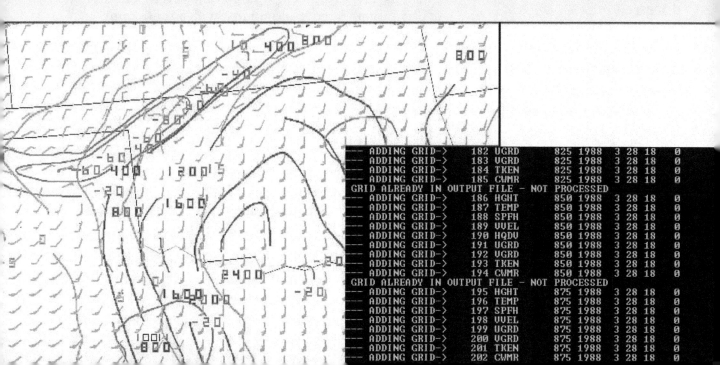

# NCAPE | Normalized CAPE

**Purpose**
- Acceleration potential of updraft parcels
- Experimental, storm severity
- Low-topped supercell environments

**Abbreviations**
NCAPE

**Introduced**
David Blanchard, 1998

**Maximized**
Very steep lapse rate above the LFC
- "Fat" positive areas

**Units**
m s$^{-2}$

**Representative values**

<0.1 - "tall and skinny" CAPE with weak parcel acceleration

0.1 to 0.3 - moderate parcel acceleration

0.3 to 0.4+ - "fat" CAPE profile with large parcel acceleration

**Examples**
| | |
|---|---|
| 0.07 | St Louis hailstorm (4/10/2001) |
| 0.09 | DFW TX / Delta 191 (8/2/1985) |
| 0.10 | Rochelle IL EF4 (4/9/2015) |
| 0.14 | Superoutbreak (KY) (4/3/1974) |
| 0.17 | Vilonia AR EF4 (4/27/2014) |
| 0.17 | Topeka KS F5 (6/8/1966) |
| 0.21 | Jarrell TX F5 (5/27/1997) |
| 0.21 | Fort Worth F3 (3/28/2000) |
| 0.24 | Moore OK EF5 (5/20/2013) |
| 0.27 | Cen. Oklahoma (5/3/1999) |
| 0.29 | Ohio derecho (6/29/2012) |
| 0.32 | Joplin MO EF5 (5/22/2011) |
| 0.35 | El Reno OK EF3 (5/31/2013) |
| 0.41 | Greensburg KS EF5 (5/4/2007) |

Normalized CAPE, or NCAPE, (Blanchard 1998) attempts to identify "CAPE density" in the lowest part of the sounding, where thunderstorm updrafts have just passed the LFC (level of free convection).

The equation for normalized CAPE is:

$$NCAPE = CAPE / Z$$

where CAPE is the buoyancy in J/kg and Z is the depth of the positive area in meters. This yields NCAPE in m s$^{-2}$.

Since the 1990s forecasters have understood that a fat appearance of the positive area in the lowest part of the sounding suggests rapid acceleration of parcels ascending past the LFC. The main contributor to this appearance is a steep, almost dry-adiabatic lapse rate just above the LFC. Under these circumstances, the temperature differential $\Delta T/\Delta z$ increases rapidly with height, contributing to significant kinetic energy in the lowest part of the sounding, near the cloud base.

Early on it was expected that NCAPE would be a valuable predictor, it does not appear that it correlates well with significant tornado risk. Still, however, there is a high likelihood of strong updrafts and severe weather in general compared to a sounding with a "tall skinny" profile. High NCAPE with the same CAPE value is likely to be a more significant indicator of general severe weather potential.

In the case of low-topped or "mini-supercells", normalized CAPE has been found to be superior to regular CAPE (Davies 1993b; Wicker and Cantrell 1996). Mini-supercells are rotating storms with smaller dimensions than normal. They are common in cold-core troughs north of the polar front in regions of high dewpoint.

Since NCAPE is based on a simple measurement of positive area depth, unusual distribution of positive area, such as temperature differential maximized at higher altitudes, may provide NCAPE values that do not have associations with severe weather that are expected.

SEVERE WEATHER INDICES 139

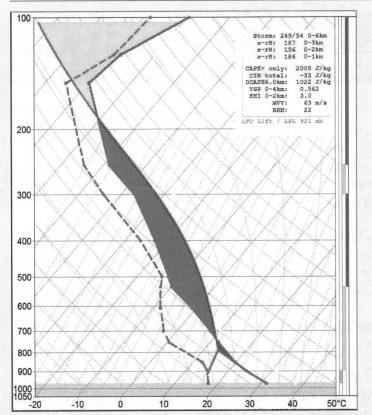

**Figure A1-5a. Low NCAPE.** Shown here is NCAPE of 0.20 m s$^{-1}$, with CAPE of 2000 J kg$^{-1}$. The positive area has a tall, skinny appearance. Calculating the depth we get about 10,000 ft depth, giving an NCAPE of 2000 / 10,000 = 0.20, verified in both RAOB and on paper. A sounding with this shape, with NCAPE in the 0.20 range, is more typical of severe weather environments than the shape below. In the summertime with high tropopause levels and low CAPE environments, NCAPE can easily fall below 0.10.

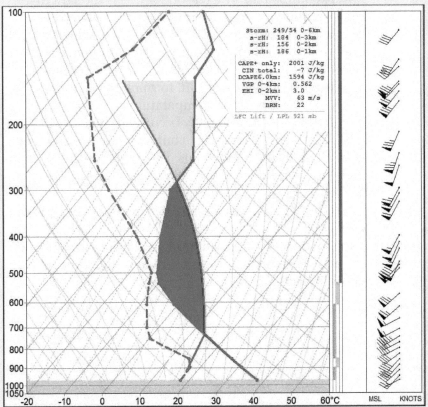

**Figure A1-5b. High NCAPE.** The NCAPE value for this sounding is 0.28, with the same CAPE quantity as the top figure. The NCAPE is based on the positive area extending from 2800 to 10,000 m (7200 m deep), which yields 2000 / 7200 = 0.28. This value agrees with the NCAPE quantity given by RAOB. Even though CAPE is still 2000 J kg$^{-1}$, the shape is "short and fat", implying parcels undergo strong accelerations just above the LFC. This shape and NCAPE value is abnormal in severe weather environments.

# LCL | Lifted condensation level

**Purpose**
Height of convective cloud base

**Abbreviations**
LCL

**Introduced**
Early 20th century

**Maximized**
Higher with dry low-level conditions

**Units**
Pressure - mb
Height - m or ft
Temperature - °C or K

**Representative values**
<1000 m      Low; night or tropical air
1000-1500 m  Typical
1500-2000 m  High
>2000 m      Very high, dry air mass

**Examples** (m AGL)
600   Vilonia AR EF4 (4/27/2014)
700   Topeka KS F5 (6/8/1966)
700   Joplin MO EF5 (5/22/2011)
800   Jarrell TX F5 (5/27/1997)
800   Cen. Oklahoma (5/3/1999)
900   Moore OK EF5 (5/20/2013)
900   El Reno OK EF3 (5/31/2013)
1100  St Louis hailstorm (4/10/2001)
1100  Fort Worth F3 (3/28/2000)
1300  Superoutbreak (KY) (4/3/1974)
1400  Rochelle IL EF4 (4/9/2015)
1400  Greensburg KS EF5 (5/4/2007)
1500  Ohio derecho (6/29/2012)
2100  DFW TX / Delta 191 (8/2/1985)

The lifted condensation level (LCL) is the level at which a parcel, forced to rise, will saturate. It may be expressed as a height, a pressure, or a temperature. Lifted condensation level is directly correlated with relative humidity; the higher the relative humidity, the lower the LCL.

The lifted parcel temperature rises along the dry adiabat, at the dry adiabatic lapse rate, until it intersects the mixing ratio corresponding to the surface dewpoint or the mean low-level mixing ratio. This is described in Chapter 3.

From that point it rises along the moist adiabat. The parcel dewpoint lifts along the mixing ratio line since mixing ratio is conserved until the parcel saturates.

The LCL has been found to have a strong association with significant tornado risk (Rasmussen and Blanchard 1998). This is maximized when the LCL is lower than 1000 m. The reason for this is thought to center around the relationship of dry air with stronger production of outflow, leading to updrafts which are undercut before supercells can develop tornadic circulations. It may also be related to dry air being entrained and ingested into the updraft, forming downdrafts with lower wet bulb potential temperature.

Tornadoes occasionally occur with higher LCLs (above 2000 m) (Davies 2006) but they are rare and tend to be short-lived. Landspouts can also be the result in high LCL conditions when boundaries and sufficient vertical vorticity are available.

The LCL will rise during the day as heating progresses. There is also usually an exchange with drier air just above the surface with moist air in contact with the surface. This mixing process further reduces the relative humidity and raises the LCL.

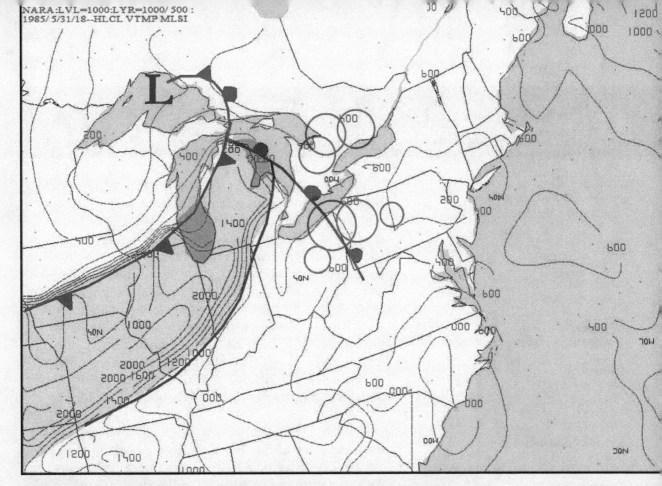

**Figure A1-6. Lifted condensation level plot (above)** (meters) for 1800 UTC, May 31, 1985, the date of the Pennsylvania-Ontario tornado outbreak. The LCL plot reveals the presence of a rare dryline or a highly modified Pacific cold front, with origins from west of the Rockies. The LCL field also reveals the distinguishes the cold front boundary, bringing more humid air from the Canadian Hudson Bay region. It does not reveal the warm front, though it is very common for LCLs poleward of a warm front to be much lower. Axes of low LCL extending equatorward tend to correlate with the deepest tropical moisture.

**Figure A1-7. Lifted condensation level (LCL)**, and its relation to the cap and a higher positive area. This air is strongly capped. *(AWS 1979)*

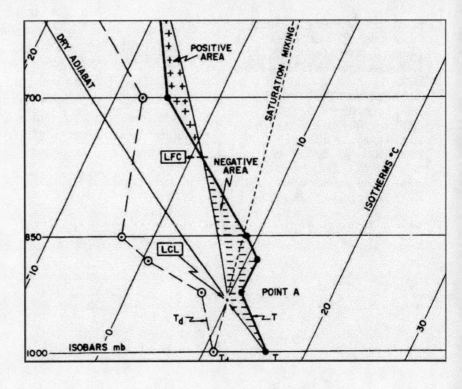

# LFC | Level of free convection

**Purpose**
- determines the base of positive area
- correlated with tornado potential

**Abbreviations**
LFC

**Introduced**
1938 or earlier

**Maximized**
Higher with dry low-level conditions

**Units**
Pressure - mb
Height - m or ft
Temperature - °C or K

**Representative values**
2000 to 3000 m

**Examples**  (m AGL)
900   Moore OK EF5 (5/20/2013)
900   Joplin MO EF5 (5/22/2011)
1000  Jarrell TX F5 (5/27/1997)
1400  Vilonia AR EF4 (4/27/2014)
1500  Greensburg KS EF5 (5/4/2007)
2000  Rochelle IL EF4 (4/9/2015)
2400  Fort Worth F3 (3/28/2000)
2500  Superoutbreak (KY) (4/3/1974)
2600  Cen. Oklahoma (5/3/1999)
2600  El Reno OK EF3 (5/31/2013)
2900  Topeka KS F5 (6/8/1966)
3000  DFW TX / Delta 191 (8/2/1985)
3000  St Louis hailstorm (4/10/2001)
3200  Ohio derecho (6/29/2012)

The level of free convection is the level at which a rising parcel becomes warmer than the surrounding air and changes from a negatively buoyant regime to a positively buoyant one.  It can be expressed as a height (m or ft), as a level (mb), or as a temperature (°C), but the most common expression is in meters or millibars.  The LFC, as with LCL, is controlled by the parcel temperature and dewpoint, as well as the lifting method (see Chapter 3).

The LFC height is inversely related to low-level moisture; that is, high amounts of low-level moisture indicate a low LFC.  The LFC height is also loosely correlated with LCL height.  A strong cap, along with warm mid-level conditions, will tend to displace the LFC to higher altitudes.

Virtual temperature should be used where possible for calculating LFC.  This accounts for density, as buoyancy is a function of density rather than temperature.  In a moist environment, the LFC will be lower than would otherwise be indicated by air temperature, as the parcel will more easily overcome negative inhibition and reach the positive region several hundred feet to a couple of thousand meters lower.

The LFC height should be less than 2500 meters (8200 ft) or lower for tornadoes to occur, though this is not necessarily a prerequisite.  The LFC height should be under 3000 meters (10,000 ft) or lower to support severe weather.

# SEVERE WEATHER INDICES 143

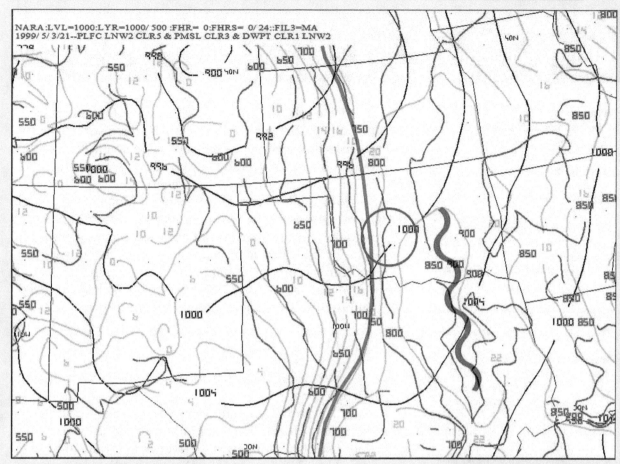

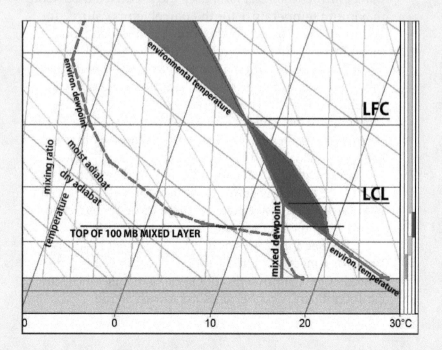

**Figure A1-8 (top). Level of free convection** (mb, red), sea level pressure (mb, black), and mixing ratio (g kg$^{-1}$, green), for May 3, 1999 at 2100 UTC, shortly before the 1999 Bridge Creek - Moore, Oklahoma tornado and nearby storms (red circle). This shows the close relationship between low LFC and high moixture (axis indicated by green wavy line).

**Figure A1-9 (left). Relationship of level of free convection (LFC)** to other features using the 100 mb mixed parcel method. The mixed dewpoint bisects the environmental dewpoint through the mixed layer. Here we see about 160 J kg$^{-1}$ of CIN (blue area) with the LFC at the top.

# LI | Lifted Index

**Purpose**
- Maximum updraft speed
- Experimental, storm severity

**Abbreviations**
LI, LIX

**Introduced**
Joseph Galway, 1956

**Maximized**
Very steep lapse rate above the LFC
- "Fat" positive areas

**Units**
m/s$^2$

**Representative values**
*(From Air Weather Service 1979)*
0 to -2   Thunderstorms possible - good trigger mechanism needed
-3 to -5  Unstable; thunderstorms probable
<-5       Very unstable - heavy to strong thunderstorm potential

**Examples**
-3    St Louis hailstorm (4/10/2001)
-3    DFW TX / Delta 191 (8/2/1985)
-4    Rochelle IL EF4 (4/9/2015)
-5    Superoutbreak (KY) (4/3/1974)
-5    Vilonia AR EF4 (4/27/2014)
-5    Topeka KS F5 (6/8/1966)
-7    El Reno OK EF3 (5/31/2013)
-8    Fort Worth F3 (3/28/2000)
-8    Cen. Oklahoma (5/3/1999)
-8    Jarrell TX F5 (5/27/1997)
-8    Moore OK EF5 (5/20/2013)
-9    Ohio derecho (6/29/2012)
-9    Greensburg KS EF5 (5/4/2007)
-11   Joplin MO EF5 (5/22/2011)

The lifted index, or LI, (Galway 1956) is a modification of the Showalter Index, using a mixed surface-based layer as the starting parcel instead of a single level at 850 mb. It was developed by Joseph Galway, one of the first Severe Local Storms Unit forecasters.

It rapidly became the benchmark index for thunderstorm forecasting between the 1960s and 1990s, and is still familiar to most severe weather forecasters as it provides a very direct, simple measurement of parcel buoyancy, and can be computed manually. The equation is identical to the Showalter index, but the initial parcel is handled differently. The index is computed as follows:

$$LI = T_{500} - T_{P500}$$

where $T_{P500}$ by definition uses a mixed layer method starting at the surface.

The parcel temperature uses either the forecast form or the static form. The *static form* uses the surface conditions exactly as shown on the sounding for the initial parcel lift. The *forecast form* uses the predicted afternoon maximum temperature for the parcel temperature. In both methods, the parcel dewpoint is obtained from the mean mixing ratio between the surface and 3000 ft AGL.

The parcel is lifted adiabatically until it reaches 500 mb. This is assumed to be the updraft temperature in a developing cloud. The difference between the updraft temperature and the surrounding environment in Celsius degrees gives the lifted index, with negative values indicating a warm, buoyant parcel.

Some studies have found that parcel lift indices like LI and SSI have low skill for tropical forecasting (Hyson et al 1964). In India, however, these indices have been found to be useful (Subramanian and Jain 1966).

Between the 1960s and 1980s, different modifications for constructing the surface parcel have been presented. One common modification is the use of the lowest 100 mb, and another uses the lowest 50 mb.

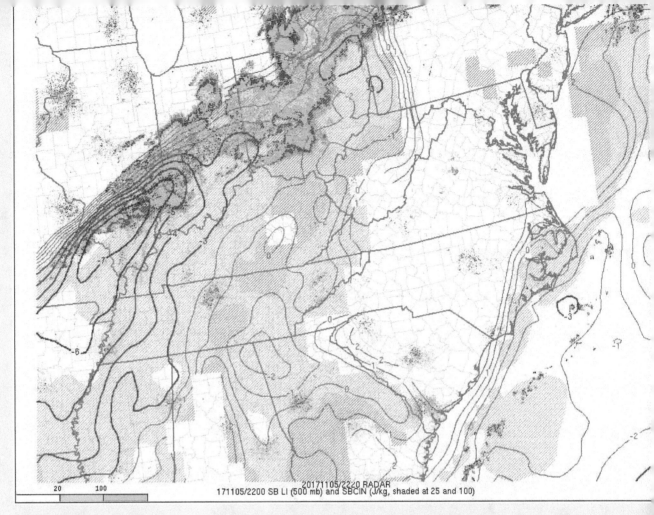

**Figure A1-10 (above): Lifted index** for November 5, 2011, showing values of -2 to -3 ahead of the bow echo region in Ohio, but as low as -7 in Illinois where discrete multicells and supercells are occurring. All CAPE values were around 500 J kg$^{-1}$ in Ohio and 1500 J kg$^{-1}$ in Illinois.

**Figure A1-11 (right): Lifted index** for the Wichita Falls F4 tornado, on April 10, 1979, 2100 UTC. Isobars are thick black lines; lifted index is thick red lines and red shading, and moisture, mixing ratio, is indicated by thin green lines. This shows a classic dryline setup just east of the Caprock, with a lifted index maximum in the Childress-Seymour area, the latter of which is where initiation took place.

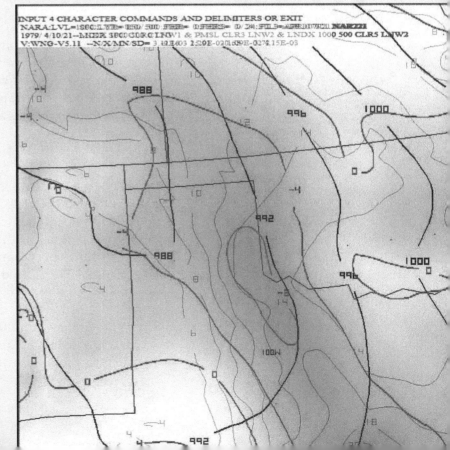

# SSI | Showalter Stability Index

**Purpose**
- Maximum updraft speed
- Experimental, storm severity

**Abbreviations**
SI, SSI

**Introduced**
Albert Showalter, 1947

**Maximized**
- Warm 850 mb temperature/dewpoint
- Cold 500 mb temperatures

**Units**
Non-dimensional; based on ΔT (°C)

**Representative values**
+3 to +1  Low chance of thunderstorms
0 to -3   Unstable / thunderstorms probable
-4 to -6  Very unstable / severe thunderstorms
<-6       Extremely unstable / severe thunderstorms with tornadoes

**Examples**
0.1   Jarrell TX F5 (5/27/1997)
−1.8  Rochelle IL EF4 (4/9/2015)
−2.1  St Louis hailstorm (4/10/2001)
−3.2  DFW TX / Delta 191 (8/2/1985)
−3.6  Vilonia AR EF4 (4/27/2014)
−3.9  Topeka KS F5 (6/8/1966)
−4.3  Superoutbreak (KY) (4/3/1974)
−5.1  Moore OK EF5 (5/20/2013)
−6.5  Cen. Oklahoma (5/3/1999)
−6.7  Fort Worth F3 (3/28/2000)
−7.3  El Reno OK EF3 (5/31/2013)
−8.8  Joplin MO EF5 (5/22/2011)
−9.4  Ohio derecho (6/29/2012)
−9.8  Greensburg KS EF5 (5/4/2007)

The SI or SSI index (Showalter 1953) was the very first index used for severe weather forecasting. It was initially developed in 1946 by Albert Showalter, a supervisor during the first few years of the Severe Local Storms Unit's existence. This index initially took the form of "Stability Index Computation Chart" and was used to forecast southern California thunderstorms. Although the index is rarely used nowadays, it is provided in this section because of this significance and its use throughout older severe storm literature.

The Showalter Index lifts a parcel from 850 mb to 500 mb. It provides a simplified estimate of buoyancy using a low-level parcel lifted into the mid-troposphere. If the parcel is warmer than the surrounding air, it is considered to be positively buoyant. It was gradually replaced by the lifted index in the 1960s and 1970s, and by CAPE shortly afterward.

The index is constructed as:

$$SI = T_{500} - T_{P500}$$

where $T_{500}$ is the environmental temperature at 500 mb and $T_{P500}$ is the temperature of a parcel lifted from 850 mb according to standard rules for lifting a parcel, therefore 850 mb temperature and dewpoint are used,

A major disadvantage of the SI index is it does not account for high terrain, yielding widely varying altitudes above ground level, often even below it in the Rockies. Therefore it was rarely used west of the Great Plains, though a modified version exists which lifts a higher level parcel, such as from 700 mb.

At least two studies found that SI values of +2 or lower were associated with severe storms in the eastern two-thirds of the US and in the Gulf Stream vicinity (David and Smith 1971) (Ellrod and Field 1984).

A modified Showalter Index known as the Curtis and Panofsky method (Curtis and Panofsky 1958) uses the mean mixing ratio of the 500-850 mb level to determine the 850 mb starting dewpoint. A method for the Colorado high plains (Hovanec and Horn 1975) substi-

SEVERE WEATHER INDICES  147

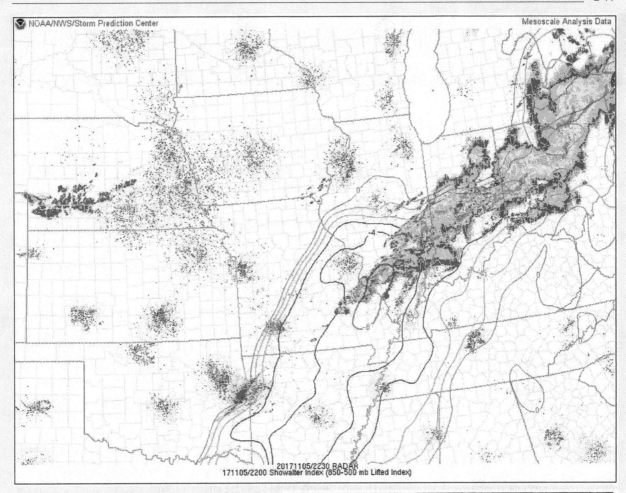

**Figure A1-12 (above): Showalter Index** for November 5, 2017. Values of –5 are found along the tail end storms in Illinois, indicating an SSI of –6 (extremely unstable).

**Figure A1-13 (right): Showalter Index** computation (AWS 1979) showing a parcel is constructed at 850 mb then lifted to 500 mb. Alternate forms of SI allow for lifting from a higher layer, such as 700 mb, or using mean mixing ratios, however this will likely provide different sets of representative values.

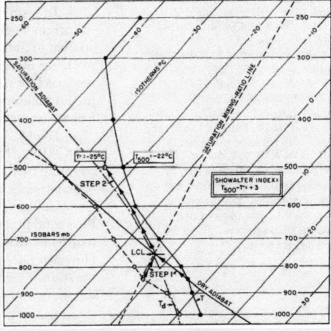

# DCAPE | Downdraft CAPE

**Purpose**
Likelihood of strong downdrafts

**Abbreviations**
DCAPE

**Introduced**
Matthew Gilmore and Louis Wicker, 1998

**Maximized**
Very dry mid-tropospheric air and warm low-level conditions

**Units**
J kg$^{-1}$

**Representative values**
<800     minimal downburst potential
800-1000 marginal downburst potential
>1000    significant for downbursts

**Examples**
0       Superoutbreak (KY) (4/3/1974)
491     St Louis hailstorm (4/10/2001)
889     Fort Worth F3 (3/28/2000)
937     Joplin MO EF5 (5/22/2011)
967     Cen. Oklahoma (5/3/1999)
1044    Vilonia AR EF4 (4/27/2014)
1097    Rochelle IL EF4 (4/9/2015)
1136    Greensburg KS EF5 (5/4/2007)
1238    Jarrell TX F5 (5/27/1997)
1262    Moore OK EF5 (5/20/2013)
1278    Topeka KS F5 (6/8/1966)
1501    El Reno OK EF3 (5/31/2013)
1625    DFW TX / Delta 191 (8/2/1985)
1967    Ohio derecho (6/29/2012)

* Note: As a reminder, the wet-bulb temperature, can be computed by locating the LCL of a parcel lifted from the original level, then descending from the LCL down the moist adiabat back to the original level. This is the wet bulb temperature, $T_w$.

Downdraft convective available of potential energy (DCAPE) (Gilmore and Wicker 1998) is an indicator of the maximum kinetic energy that could result due to evaporative cooling as a parcel descends. A descending parcel always descends dry-adiabatically, resulting in strong warming. However with DCAPE we *assume* that the parcel is replenished with rain, allowing it to sink at the moist-adiabatic lapse rate. In a few storms with very high precipitation, the actual realized energy could be higher than this due to precipitation loading.

To compute DCAPE graphically, it is necessary to have a third line showing wet bulb temperature*. This can be plotted with tools such as RAOB. The downdraft origin level is the level between the surface and 400 mb with the lowest mean wet bulb temperature in a layer 100 mb thick. With this, obtain the downdraft origin level and the downdraft origin wet-bulb temperature.

An alternate method is to use the 600 mb wet bulb temperature, the the 700 mb wet bulb temperature, the mean 500-700 mb level and wet bulb temperature, or the level with the driest $T_w$ between the surface and 400 mb (Rasmussen 1994) as the origin level.

To form the left side of the polygon, descend from the origin level's wet bulb temperature down the moist adiabat to the surface. The right side of the polygon is made up of the temperature profile, $T$. This should be modified for conditions at the time of convection. The top and bottom of the polygon is the 600 mb and surface isobars connecting the left and right side polygons.

The DCAPE polygon's geometric area is equivalent to DCAPE. This can be converted to J kg$^{-1}$ using the box counting method, or a J kg$^{-1}$ value can be estimated based on experience.

Unlike CAPE, DCAPE does not attempt to model the actual process in the downdraft and cannot be used to directly calcualte downdraft temperature or velocity. However DCAPE is related to the likelihood of strong downdraft velocity, and larger values imply greater downdraft strength.

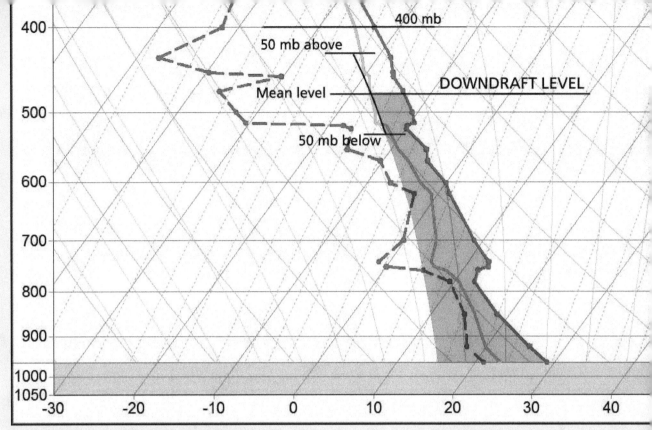

**Figure A1-14 (above): Computation of DCAPE** as observed at Oklahoma City, hours ahead of a derecho, the "People Chaser" event of May 27-28, 2001. The center line is the wet bulb profile and the purple shading is the DCAPE polygon. The calculated value for this sounding was 1383 J kg$^{-1}$

**Figure A1-15 (below): DCAPE** from the SPC Experimental Mesoanalysis page on November 5, 2017. Scattered warnings around the Cleveland and Dayton area forecast gusts of 60-70 mph.

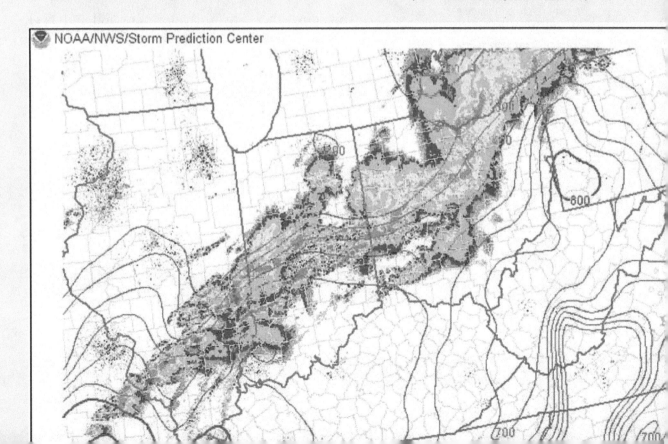

# KI | K Index

**Purpose**
- Quick areal thunderstorm forecasting

**Abbreviations**
K, KI

**Introduced**
R. M. Whiting, late 1950s
Published by Joseph J. George, 1960

**Maximized**
- Steep low-mid level lapse rates
- Rich low-level water vapor content
- Moist mid-level relative humidity

**Units**
None, but yields a °C value

**Representative values**
*(from George, 1960)*
*Subtract 5 from values below for US locations west of the Rockies (AWS 1979)*
<20      None
20-25    Isolated thunderstorms
26-30    Widely scattered thunderstorms
31-35    Scattered thunderstorms
>35      Numerous thunderstorms

**Examples**
20.5   Jarrell TX F5 (5/27/1997)
25.3   Fort Worth F3 (3/28/2000)
25.5   Rochelle IL EF4 (4/9/2015)
26.9   Moore OK EF5 (5/20/2013)
27.0   St Louis hailstorm (4/10/2001)
27.4   Cen. Oklahoma (5/3/1999)
28.0   Topeka KS F5 (6/8/1966)
32.2   Joplin MO EF5 (5/22/2011)
32.9   Ohio derecho (6/29/2012)
33.3   El Reno OK EF3 (5/31/2013)
34.3   Superoutbreak (KY) (4/3/1974)
37.5   DFW TX / Delta 191 (8/2/1985)
39.3   Vilonia AR EF4 (4/27/2014)
44.3   Greensburg KS EF5 (5/4/2007)

The K Index is credited to R. M. Whiting of the Eastern Air Lines Meteorological Department in Atlanta in a compendium of forecasting techniques published by the department's director (George 1960). It was intended to solve the problem of predicting thunderstorm activity, specifically with non-frontal storms occurring with relatively weak winds aloft. It was developed specifically for areal thunderstorm forecasting, where an extensive evaluation of thunderstorm potential at each station was not possible. This allowed forecasters to generate accurate briefing maps covering entire continents.

At the time, the importance of positive area, or CAPE, on soundings were well-understood, but the lack of computing technology meant that analysts needed a technique for quickly summarizing values, station by station, based on radiosonde data.

The K Index is defined as:

$$KI = (T_{850} - T_{500}) + T_{d850} - (T_{700} - T_{d700})$$

This shows there are three terms: low and mid-level lapse rate; moisture content of the lower atmosphere; and mid-level relative humidity, suggesting moisture depth.

The first term $(T_{850} - T_{500})$ considers the lapse rate, and is essentially the same as the vertical totals index. The greater this term, the greater the probability that a rising parcel will continue to rise.

The second term $T_{d850}$ takes into account the low-level moisture by using the 850 mb dewpoint.

The third term $(T_{700} - T_{d700})$ examines the dewpoint depression at 700 mb. Therefore, dry 700 mb conditions with low relative humidity reduces K Index, while high values increase it. This is the Achilles hell of the K Index, as most Great Plains severe weather outbreaks involve a very dry layer above 850 mb. Therefore it tends to fail in the central and western states, but provides good results for deep, warm season southeast flow in the eastern U.S..

A study during the 1970s found that it was the best available index for non-severe summertime thunderstorms. However the same study found it to be a poor predictor of severe weather.

SEVERE WEATHER INDICES  151

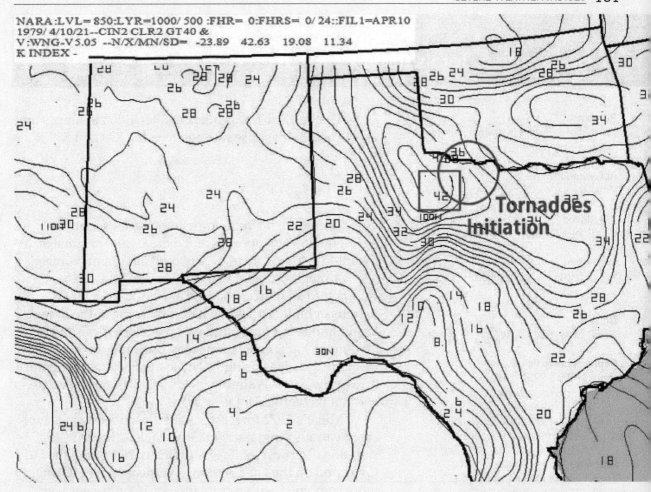

**Figure A1-16 (above): K Index plot** for 2100 UTC on April 10, 1979, as the storms developed that would later produce the Wichita Falls, Texas F4 tornado. *(WinGRIDDS plot)*

**Figure A1-17 (right): K Index product** from the Climate Prediction Center covering the US. This shows the higher instability observed in a barolinic frontal system off the Atlantic coast in October 2017.

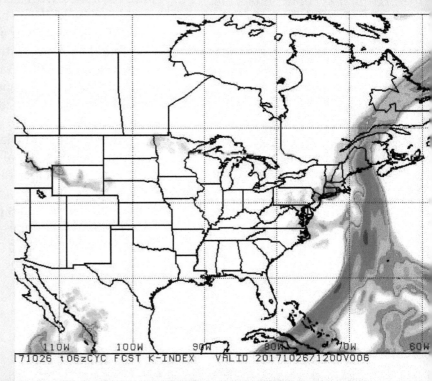

# VT | Vertical Totals Index

**Purpose**
- Identifies a favorable environment for buoyant parcels

**Abbreviations**
VT

**Introduced**
Col. Robert C. Miller, 1967

**Maximized**
Very steep lapse rate between 850 and 500 mb

**Units**
None, but yields a °C value

**Representative values**
*(from Miller 1972)*
<28     No thunderstorms
29-32   Few thunderstorms
>32     Scattered thunderstorms
40-43   Approximate dry adiabatic limit

**Examples**
26.1   Vilonia AR EF4 (4/27/2014)
27.3   DFW TX / Delta 191 (8/2/1985)
27.7   El Reno OK EF3 (5/31/2013)
27.7   Topeka KS F5 (6/8/1966)
28.2   St Louis hailstorm (4/10/2001)
28.3   Moore OK EF5 (5/20/2013)
29.3   Rochelle IL EF4 (4/9/2015)
30.5   Superoutbreak (KY) (4/3/1974)
30.9   Joplin MO EF5 (5/22/2011)
31.3   Ohio derecho (6/29/2012)
31.9   Cen. Oklahoma (5/3/1999)
31.9   Fort Worth F3 (3/28/2000)
31.9   Jarrell TX F5 (5/27/1997)
34.9   Greensburg KS EF5 (5/4/2007)

The Vertical Totals index (Miller 1967) is simply an expression of the lapse rate between the 850 and 500 mb levels:

$$VT = T_{850} - T_{500}$$

Therefore, it simply measures the lapse rate between the lower troposphere, where inflow originates, to the middle troposphere, where buoyancy is maximized.

A major problem with the vertical totals index is common to all indices that use a fixed-layer method: regions of high terrain may intersect some of the index's levels, and bring the boundary layer close these levels. Stations in higher elevations may use a modification of this index, such as using 700 instead of 850 mb temperatures, however this simply yields 700-500 mb lapse rate and not the vertical totals index.

Miller's 1972 forecaster notes described values typical of various air masses: in Great Plains "loaded gun" situations, vertical totals was typically 28. Along the Gulf Coast and on the Gulf Stream, a vertical total of 23 or more combined with a cross total of 16 or greater were considered by Miller to be adequate for storms.

VT values of 26 or above are typically associated with thunderstorms, with 30 or above indicating severe convection. The 6.7° C/km lapse rate that comprises a threshold value for tornado outbreaks (Craven 2000) corresponds to a VT of 28 to 29 in most severe weather environments.

The dry adiabatic lapse rate provides an upper boundary on the Vertical Totals value, as layers exceeding the dry adiabatic lapse rate will normally mix out This limit ranges from 37 in a winter air mass (theta of 0°C) to 43 in a summer air mass (theta of 40°C). The AMS Glossary (AMS, 2017) defines approximate values of 40 as the dry-adiabatic lapse rate; and 20 to 30 for the moist adiabatic lapse rate when the 850 mb temperature is 15° and 0°C, respectively.

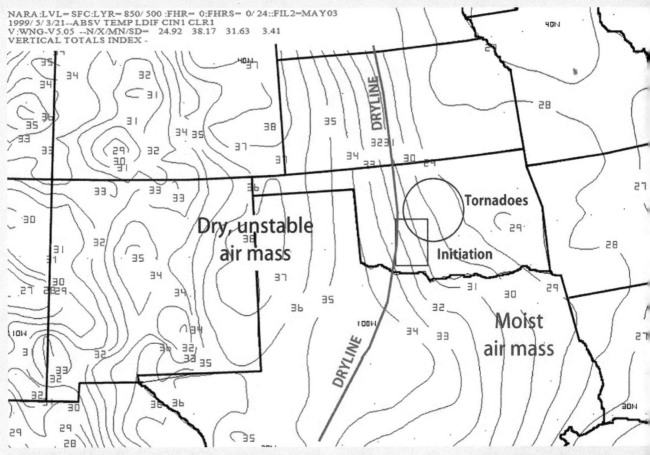

**Figure A1-18 (above): Vertical totals** plot for May 3, 1999 at 2100 UTC, as large supercells were forming west and southwest of Oklahoma City, eventually producing 72 tornadoes including the F5 Moore, Oklahoma tornado. This shows the extreme lapse rates that are common in the warm, strongly mixed air west of the dryline.

**Figure A1-19 (left): Colonel Robert C. Miller**, creator of the early severe weather forecasting indices (center front) alongside **Ernest J. Fawbush** at right. Behind them at center right is Major **L. J. Starrett**, another Air Force weather officer who helped write some of the presentations on these methods.

# CT | Cross Totals Index

**Purpose**
- Maximum updraft speed
- Experimental, storm severity

**Abbreviations**
CT

**Introduced**
Col. Robert C. Miller, 1967

**Maximized**
- High dewpoint at 850 mb
- Cold temperatures at 500 mb

**Units**
None, but yields a °C value

**Representative values**
<18    Thunderstorms unlikely
18-29  Thunderstorms
≥30    Severe thunderstorms

**Examples**
18.0   Jarrell TX F5 (5/27/1997)
22.3   DFW TX / Delta 191 (8/2/1985)
23.3   Rochelle IL EF4 (4/9/2015)
23.9   St Louis hailstorm (4/10/2001)
24.0   Topeka KS F5 (6/8/1966)
24.7   Vilonia AR EF4 (4/27/2014)
25.0   Superoutbreak (KY) (4/3/1974)
26.3   Moore OK EF5 (5/20/2013)
26.7   El Reno OK EF3 (5/31/2013)
27.4   Cen. Oklahoma (5/3/1999)
27.9   Fort Worth F3 (3/28/2000)
28.1   Greensburg KS EF5 (5/4/2007)
28.5   Ohio derecho (6/29/2012)
29.6   Joplin MO EF5 (5/22/2011)

The Cross Totals Index (Miller 1967) is a companion to the Vertical Totals index. It simply compares the low-level moisture to the mid-tropospheric temperature.

$$CT = T_{d850} - T_{500}$$

The index is constructed so that the value is maximized in the presence of rich low-level moisture and cold mid-tropospheric temperatures, the latter of which implies steep lapse rates.

The first term $T_{d850}$ establishes the amount of low-level moisture. The higher the dewpoint, the higher the cross totals index is. This measurement is common in many other indices; it makes up the derivative total totals index as well as the K Index.

The second term quantifies the mid-tropospheric temperatures; the lower the temperature, the higher the result. By relating it to the first term, we see that very moist 850 mb air and very cold 500 mb air maximize the cross total index.

The main problem of cross totals is the use of fixed levels. This renders it unusable in higher terrain, such as that in the western half of the United States. Furthermore, the cross totals index is critically dependent on whether moisture is deep enough to reach 850 mb. In many springtime events, moisture may be up to 5000 feet deep, while the 850 mb level resides within the dry capping inversion. In this case the cross totals index will underestimate thunderstorm potential.

A CT value of 18 or greater is associated with a lower limit for thunderstorm activity. Miller's 1967 and 1972 papers identified this value as a first guess for thunderstorm activity, but prescribed 16 as adequate for thunderstorms along the Gulf Coast and on the Gulf Stream when coupled with a vertical total of 23 or greater. Cross total values of 30 or greater are associated with severe activity.

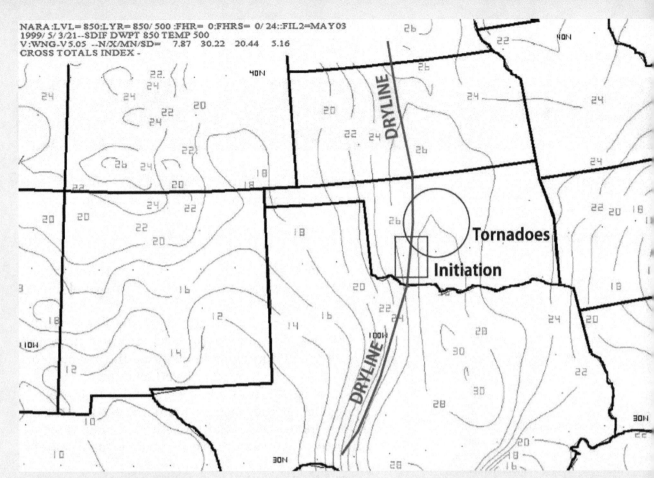

**Figure A1-20: Cross Totals Index** for May 3, 1999 at 2100 UTC, similar to the graphic on the previous page. This shows the sensitivity to low-level moisture, while in the moist sector it gives some indication of the lapse rate although it is not directly measured.

**Figure X-XX:** Col Robert C. Miller, the creator of some of the early thermodynamic indices.

# TTI | Total Totals Index

**Purpose**
- Maximum updraft speed
- Experimental, storm severity

**Abbreviations**
TT, TTI, TOTL

**Introduced**
Col. Robert C. Miller, 1967

**Maximized**
Very steep lapse rate above the LFC
- "Fat" positive areas

**Units**
m/s$^2$

**Representative values**
44-50   General thunderstorms
51-55   Moderate thunderstorms
55-59   Strong thunderstorms
60+     Scattered severe and tornadoes

**Examples**
49.6   DFW TX / Delta 191 (8/2/1985)
49.9   Jarrell TX F5 (5/27/1997)
50.8   Vilonia AR EF4 (4/27/2014)
51.7   Topeka KS F5 (6/8/1966)
52.1   St Louis hailstorm (4/10/2001)
52.6   Rochelle IL EF4 (4/9/2015)
54.4   El Reno OK EF3 (5/31/2013)
54.6   Moore OK EF5 (5/20/2013)
55.5   Superoutbreak (KY) (4/3/1974)
59.3   Cen. Oklahoma (5/3/1999)
59.8   Fort Worth F3 (3/28/2000)
59.8   Ohio derecho (6/29/2012)
60.5   Joplin MO EF5 (5/22/2011)
63.0   Greensburg KS EF5 (5/4/2007)

The Total Totals Index (Miller 1967) combines the Vertical Totals and Cross Totals Index, as follows:

$$TT = VT + CT = T_{850} + T_{d850} - 2T_{500}$$

The total totals index is similar to the K Index developed by Eastern Air Lines, and is comprised of three primary elements.

The first term $T_{850}$ provides a baseline of the energy available near the storm inflow level. Temperature at 850 mb is related to warm air advection and the potential for buoyancy given constant mid-level conditions.

Likewise $T_{d850}$ is the other element that factors in the potential for buoyancy by examining dewpoint near the storm inflow level. Moisture is actually four times as significant as temperature in achieving the same amount of buoyancy.

The final term $T_{500}$ summarizes the mid-tropospheric temperatures, providing a measure of the possibility for buoyancy of parcels. The colder the temperature, the higher the resulting total totals index. This also carries a weighting coefficient of 2, so cold temperatures at this level are especially effective at influencing the final result. These cold mid-level temperatures are normally associated with strong dynamic lift which promote widespread cap removal, so it can be inferred that the total totals index is effective at predicting mesoscale convective systems (squall lines). Indeed, values of 50 or above have been used for forecasting mesoscale convective systems (Rodgers et al 1984).

Miller's original notes stated 44 or above as the first-guess threshold for thunderstorm activity, with or higher are associated with thunderstorms, with 60 associated with scattered severe thunderstorms and tornadoes (Miller 1975).

Miller's 1972 forecaster notes described total totals values typical of various air masses: in Great Plains "loaded gun" situations, totals totals was typically 54; in Gulf Coast tropical air masses, it was also 54; in cold

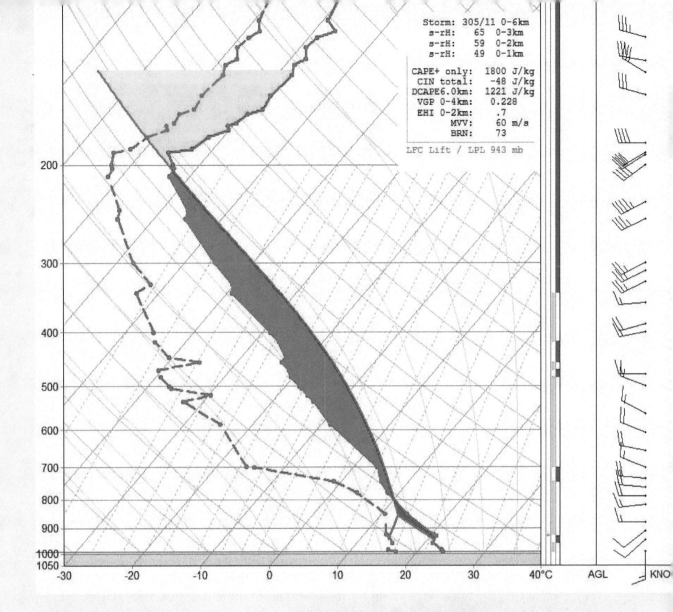

**Figure A1-21: Sounding with very high Total Totals Index**, equaling 60.1. The warm temperatures and high dewpoint at 850 mb, contrasted with very cold 500 mb conditions, -18°C, provide high numbers for this index.

**Figure A1-22: The Total Totals Index for South America** is part of the Climate Prediction Center suite of short-term products.

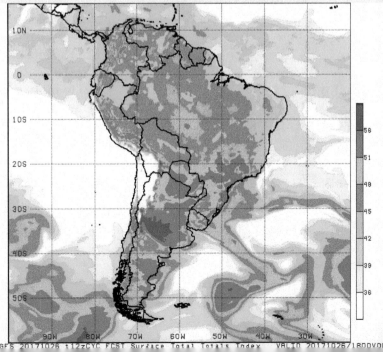

# SRH | Storm relative helicity

**Purpose**
Identifies environments that support rotating updrafts

**Introduced**
Douglas Lilly, 1986

**Maximized**
Strong directional change and magnitude through a layer relative to an updraft

**Units**
$m^2\ s^{-2}$

**Representative values**

0-3 km
>250   Supercells with tornadoes

0-1 km
>100 Supercells with tornadoes

**Examples**
(first number is 0-3 km, second number is 0-1 km AGL)

0-3 km
| | |
|---|---|
| 56 | Jarrell TX F5 (5/27/1997) |
| 101 | Fort Worth F3 (3/28/2000) |
| 107 | Ohio derecho (6/29/2012) |
| 179 | Greensburg KS EF5 (5/4/2007) |
| 213 | Moore OK EF5 (5/20/2013) |
| 217 | St Louis hailstorm (4/10/2001) |
| 226 | Joplin MO EF5 (5/22/2011) |
| 406 | Cen. Oklahoma (5/3/1999) |
| 468 | Rochelle IL EF4 (4/9/2015) |
| 480 | El Reno OK EF3 (5/31/2013) |
| 545 | Topeka KS F5 (6/8/1966) |
| 578 | Superoutbreak (KY) (4/3/1974) |
| 780 | Vilonia AR EF4 (4/27/2014) |

0-1 km
| | |
|---|---|
| -25 | Jarrell TX F5 (5/27/1997) |
| 45 | Ohio derecho (6/29/2012) |
| 69 | Greensburg KS EF5 (5/4/2007) |
| 142 | Moore OK EF5 (5/20/2013) |
| 144 | Fort Worth F3 (3/28/2000) |
| 151 | St Louis hailstorm (4/10/2001) |
| 201 | Joplin MO EF5 (5/22/2011) |
| 281 | El Reno OK EF3 (5/31/2013) |
| 299 | Cen. Oklahoma (5/3/1999) |
| 315 | Topeka KS F5 (6/8/1966) |
| 318 | Rochelle IL EF4 (4/9/2015) |
| 391 | Superoutbreak (KY) (4/3/1974) |
| 485 | Vilonia AR EF4 (4/27/2014) |

Storm relative helicity, SRH, reflects the streamwise vorticity in the storm inflow layer. It is maximized when radiosonde data shows strong directional change in the lowest few kilometers of the troposphere, along with moderate to high wind velocities. This products an elongated, curved hodograph shape.

Storm-relative helicity measures the potential for rotating updrafts. It is one of the most reliable standalone indicators among all indices at discriminating between non-supercells, nontornadic supercells, and tornadic supercells.

Storm-relative helicity is not considered through the entire depth of the storm, but through a shallow surface-based low-level layer representing the storm inflow layer. This layer was originally defined to be 0-3 km in the 1990s, as storm modeling favored the use of deep storm inflow layers. Later work identified the 0-1 km layer as especially significant for tornado forecasting, and it is believed that many shear calculations should focus on this layer. Potential problems are that such fine-grained measurements of SRH require accurate surface winds which are representative of the inflow parcel. Also conventional TEMP (TTAA-TTBB-PPBB) has limited resolution and has a high threshold before more measurement datapoints have to be added within a layer. The new high-resolution BUFR upper air format is partially designed to rectify problems like this.

The effective inflow layer storm relative helicity (ESRH) method has been developed (Thompson et al 2007) to more properly handle deeper inflow layers or layers which are elevated, a common occurrence during the cold season. It can be thought of as a "floating" storm inflow layer, which adheres to the lowest layers where significant buoyancy is found. This is not necessarily based at the surface. Values of ESRH below 60 are associated with non-supercell storms, with values of above 100-150 associated with tornadic supercells. Significant tornadoes show a mean ESRH of about 240.

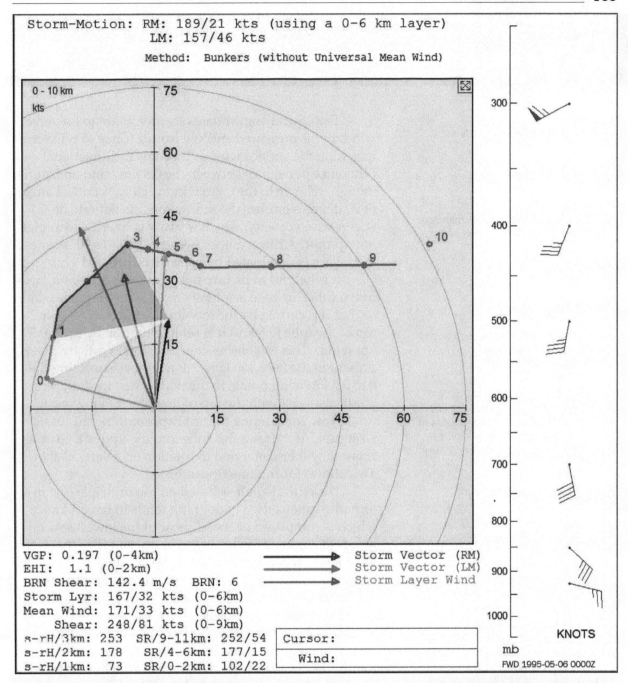

**Figure A1-23. Storm-relative helicity (SRH)** as depicted graphically, with profile wind height in km (red). The polygon representing the total area is formed at the storm vector, normally the deviant right-motion vector (pictured here), but can be assigned to whichever one is representative. A line is drawn to the 0 km wind vector and another is drawn to either the 1 km or 3 km wind vector. The wind profile line (red) provides the opposite side. Then the area within the polygon is measured. If we are just considering 0-1 km SRH, we use only the yellow polygon shown above. If we are considering 0-3 km SRH, we use both the yellow and green polygons. The hodograph shows strong northward movement on cells; the wind barbs at right show this because of a combination of strong easterly low-level flow and modest upper-level flow.

# CA | Critical angle

**Purpose**
Measure of streamwise vorticity

**Introduced**
John Esterheld & Donald Giuliano, 2008

**Maximized**
90°; 0-1 km shear vector is perpendicular to 0 km storm-relative wind vector

**Units**
degrees

**Representative values**

| | |
|---|---|
| 90 | Maximum streamwise vorticity |
| <100 | Tornadic supercells |
| >100 | Non-tornadic storms |

**Examples**
*These are guideline values as may be initially observed in sounding data, and are not necessarily based on profiles that actually existed in the inflow area of these storms.*

| | |
|---|---|
| 49 | Topeka KS F5 (6/8/1966) |
| 49 | DFW TX / Delta 191 (8/2/1985) |
| 53 | Rochelle IL EF4 (4/9/2015) |
| 56 | Jarrell TX F5 (5/27/1997) |
| 57 | Vilonia AR EF4 (4/27/2014) |
| 60 | Superoutbreak (KY) (4/3/1974) |
| 67 | Joplin MO EF5 (5/22/2011) |
| 70 | Ohio derecho (6/29/2012) |
| 75 | Moore OK EF5 (5/20/2013) |
| 87 | Fort Worth F3 (3/28/2000) |
| 87 | Greensburg KS EF5 (5/4/2007) |
| 89 | St Louis hailstorm (4/10/2001) |
| 97 | Cen. Oklahoma (5/3/1999) |
| 113 | El Reno OK EF3 (5/31/2013) |

The critical angle relates storm motion to low-level wind, but is measured entirely from a frame of reference matching the surface winds. It equals the directional difference in degrees between the 0.5 km wind and storm motion, relative to the 0 km wind vector. A critical angle of 90 degrees implies the inflow consists entirely of streamwise vorticity, which is easily converted to a rotating updraft. Critical angles perpendicular to 90° suggest the inflow is dominated by crosswise vorticity.

It is helpful to picture horizontal vorticity as a cardboard tube. Its base is initially fixed with the 0 km wind, so that it is carried by the low-level wind flow. At the top of the tube (0.5 km) it is set into motion by the 0-0.5 km wind. This establishes the horizontal vorticity that is present in the 0-0.5 km layer. If a vector connecting the 0 to 0.5 km wind points to the north, then the tube will point east-west with the top of the tube moving north.

Now considering a storm, approaching this east-west tube. If it draws this tube into the updraft, all of its rotation will be converted to rotation on a vertical axis. This allows for rotating updrafts.

The critical angle is based on the principle that in tornadic supercells, a hodograph tends to have a kink which is comprised of speed shear in the lowest several hundred meters, as velocity increases, and then this is followed by directional shear at higher levels. This provides the initial extension of the hodograph toward the upper-level wind vectors. Both of these together provide for elevated storm-relative helicity values.

In the seminal paper by Esterheld and Giuliano on critical angle which examined 67 severe storms, it was found that the storms in their dataset which were tornadic showed critical angles of about 90, while nonsevere storms showed angles of about 110°.

It should be noted that as with most shear-related measures, CA depends on an accurate estimate of near-storm wind profiles to have meaningful information. In the example values at left, no major effort was made to calculate

SEVERE WEATHER INDICES 161

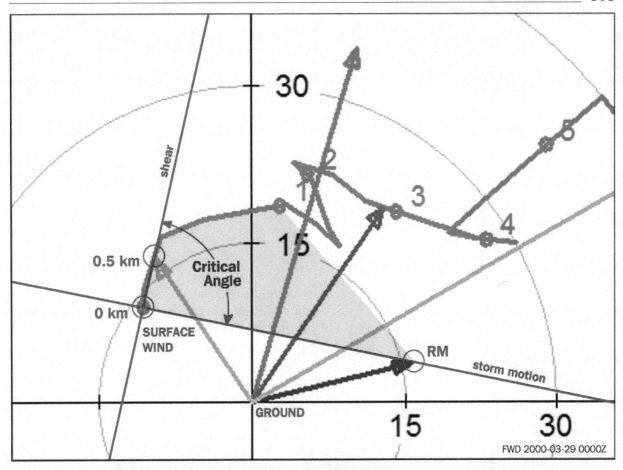

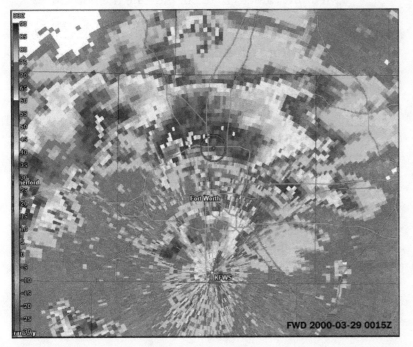

**Figure A1-24 (above). Critical angle example** showing how this parameter is worked out. The 0-1 km storm-relatve helicity is shaded in dark yellow. This shows that the shear line is closely related to one of the edges that forms the SRH polygon, but departures from this relationship will be observed if there is directional shear in the 0-0.5 km layer. Event is the Fort Worth 3/28/00 tornado (FWD). This yields CA = 87°.

**Figure A1-25 (left).** Corresponding radar image, two minutes after time of radiosonde launch (0013Z) with the launch site circled. This shows an excellent proximity sounding, although the site is close to or in the forward flank downdraft. There was likely some rain and hail coming down at the time.

# BRN | Bulk Richardson Number

**Purpose**
Showing balance of buoyancy and shear

**Abbreviations**
BRN

**Introduced**
Moncrieff and Green, 1972 (Ri)
Weisman and Klemp, 1982 (BRN)

**Maximized**
High CAPE and low shear (common in summer)

**Units**
m/s$^2$

**Representative values**

| | |
|---|---|
| <10 | Weak instability or strong shear |
| 10-30 | Supercells |
| 30-40 | Supercells; unsteady multicells |
| 40-50 | Unsteady multicells |
| >50 | Outflow dominant storm |

**Examples - BRN**

| | |
|---|---|
| 5 | Superoutbreak (KY) (4/3/1974) |
| 6 | Rochelle IL EF4 (4/9/2015) |
| 8 | St Louis hailstorm (4/10/2001) |
| 10 | Vilonia AR EF4 (4/27/2014) |
| 13 | Topeka KS F5 (6/8/1966) |
| 15 | El Reno OK EF3 (5/31/2013) |
| 18 | Moore OK EF5 (5/20/2013) |
| 21 | Cen. Oklahoma (5/3/1999) |
| 21 | Fort Worth F3 (3/28/2000) |
| 27 | Jarrell TX F5 (5/27/1997) |
| 37 | Ohio derecho (6/29/2012) |
| 42 | Joplin MO EF5 (5/22/2011) |
| 104 | Greensburg KS EF5 (5/4/2007) |
| 441 | DFW TX / Delta 191 (8/2/1985) |

**Examples - BRN shear (0.5U$^2$)**

| | |
|---|---|
| 1.9 | DFW TX / Delta 191 (8/2/1985) |
| 47.2 | Greensburg KS EF5 (5/4/2007) |
| 77.3 | St Louis hailstorm (4/10/2001) |
| 80.9 | Ohio derecho (6/29/2012) |
| 84.6 | Fort Worth F3 (3/28/2000) |
| 92.1 | Joplin MO EF5 (5/22/2011) |
| 93.6 | Jarrell TX F5 (5/27/1997) |
| 110 | Cen. Oklahoma (5/3/1999) |
| 116 | Topeka KS F5 (6/8/1966) |
| 143 | Moore OK EF5 (5/20/2013) |
| 143 | Rochelle IL EF4 (4/9/2015) |
| 195 | El Reno OK EF3 (5/31/2013) |
| 210 | Vilonia AR EF4 (4/27/2014) |
| 232 | Superoutbreak (KY) (4/3/1974) |

The BRN, dimensionless, expresses the ratio between work done by buoyancy versus by wind shear, and attempts to determine the balance between inflow and outflow. It uses the 0-6 km bulk shear magnitude. Low numbers indicate shear is dominant, while high numbers indicate buoyancy is dominant. The equation is as follows:

$$R = B / 0.5\ U^2$$

where $B$ equals buoyancy, given by mixed-layer CAPE, J kg$^{-1}$, and $U$ equals the vector difference between the 0-0.5 km mean wind and the 5.5-6.0 km mean wind, in m s$^{-1}$, thus $U$ is roughly equivalent to the 0-6 km bulk shear. The wind is mass-weighted.

Low BRN values (below 10) are associated with weak instability, strong shear, or both. Towering cumulus, "turkey towers", and developing storms will tend to shear apart, and there is a likelihood that storms will be unable to form due to the strong shear.

Medium values (10 to 50) indicate the buoyancy and shear are balanced, and this is the most conducive to steady-state storms: supercells. This medium range of values favoring supercells is based on early numerical modeling.

High BRN values (above 50) are associated with strong instability, weak shear, or both. Single cells or multicell clusters are generally favored. However if $R$ is too large, the inflow may not be strong enough to supply the updraft given the available instability, and storms will have difficulty reaching a steady state mode.

BRN is closely related to energy helicity index (EHI), which is a product rather than a ratio, and considers storm-relative helicity instead of bulk shear. The EHI parameter is more appropriate for forecasting rotating storms and tornadoes, while BRN indicates the potential for organized, long-lived storms. Also *BRN is a ratio* and can indicate favorable supercell conditions when CAPE and shear are marginal, so it must be used in conjunction with other forecast techniques.

SEVERE WEATHER INDICES 163

**Figure A1-26: Bulk Richardson number spectrum** as given in the seminal paper on BRN by M. Weisman and J. Klemp, building on the work of Moncrieff and Green and laying groundwork for the early storm simulations. "Secondary storms" refer to development of new cells after the initial updraft and downdraft pair. *(Weisman and Klemp 1982)*

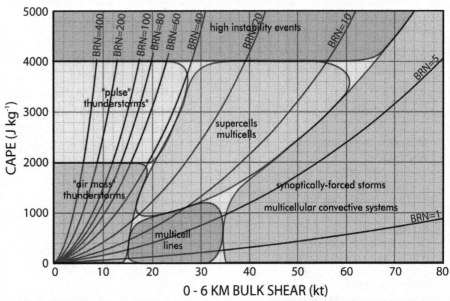

**Figure A1-27: Bulk Richardson parameter space** with shear on the X-axis and buoyancy on the Y-axis. An overprint of BRN (curved lines) is shown. This illustrates how high BRN is associated with weak shear and low BRN is associated with weak instability. Typical storm types are indicated by shading.

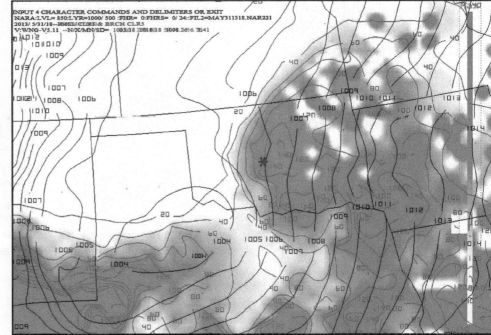

**Figure A1-28: Bulk Richardson Number** plot for May 31, 2013 at 1800 UTC showing isobars (black) and BRN (red isopleths and red shading) a few hours before development of the El Reno tornado supercell. It appears BRN was useful in showing the favorable pre-storm environment in the Oklahoma City area.

# VGP | Vorticity Generation Parameter

**Purpose**
Relates shear to buoyancy to assess the potential for rotating updrafts

**Abbreviations**
VGP

**Introduced**
Erik Rasmussen and David Blanchard, 1998

**Maximized**
High instability and high deep-layer shear

**Units**
m s$^{-2}$

**Representative values**
(From RAOB)
<0.15   Non-tornadic storms
0.15-0.25   Threshold values
>0.25   Tornadic storms

**Examples**
0.08   DFW TX / Delta 191 (8/2/1985)
0.20   St Louis hailstorm (4/10/2001)
0.31   Rochelle IL EF4 (4/9/2015)
0.34   Fort Worth F3 (3/28/2000)
0.38   Topeka KS F5 (6/8/1966)
0.40   Superoutbreak (KY) (4/3/1974)
0.41   Greensburg KS EF5 (5/4/2007)
0.46   Cen. Oklahoma (5/3/1999)
0.54   Moore OK EF5 (5/20/2013)
0.56   Jarrell TX F5 (5/27/1997)
0.56   Ohio derecho (6/29/2012)
0.59   Vilonia AR EF4 (4/27/2014)
0.80   El Reno OK EF3 (5/31/2013)
0.87   Joplin MO F5 (2011-05-22)

The vorticity generation parameter attempts to estimate the rate of conversion of horizontal vorticity (rotation along a horizontal axis, like a pencil lying on a desk) to vertical vorticity (like an updraft or tornado), through tilting and stretching, by a thunderstorm updraft (Rasmussen and Blanchard 1998). It equals:

$$VGP = S \cdot (CAPE)^{1/2}$$

where $S$ as originally defined equals the mean shear, determined by the hodograph length divided by its depth, and $(CAPE)^{1/2}$. This last term is the same as the square root of the CAPE, thus increases in CAPE contribute more at lower CAPE values than at higher ones.

The index is similar to EHI but considers shear instead of storm-relative helicity. The original definition used the 0-4 km AGL shear layer, but 0-3 km is now typically used.

Values of 0.20 m s$^{-2}$ are considered to be the threshold value for tornadic storms. Significant tornado events tend to fall within the range of 0.35 to 0.55, with outliers due to high shear or CAPE contributions.

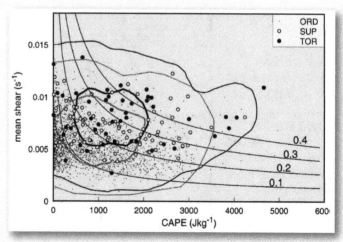

**Figure A1-29. Vorticity generation parameter** relation to storm events (plotted), CAPE (X-axis), mean shear (Y-axis), and VGP (curves). *(From Rasmussen & Blanchard 1998)*

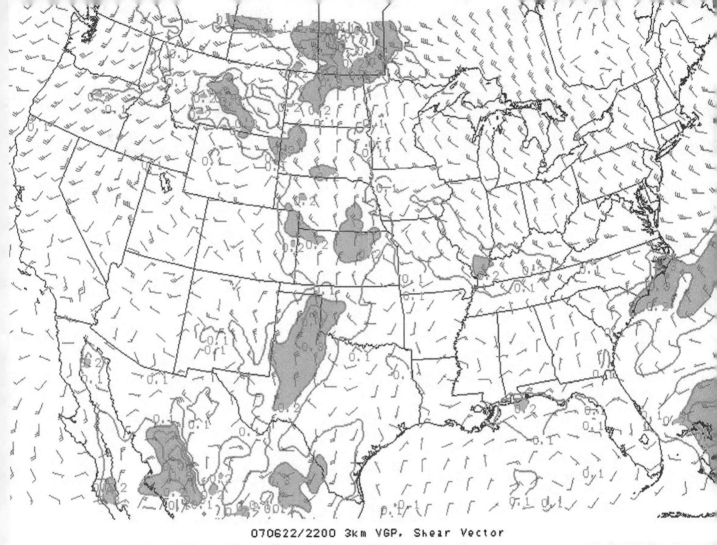

070622/2200 3km VGP, Shear Vector

**Figure A1-30 (above): Vorticity generation parameter** for 5 pm CDT on June 22, 2007, the date of the Elie, Manitoba F5 tornado. Values of about 0.3 are indicated in the area, consistent with tornadic storms. *(SPC)*

**Figure A1-31 (right): Vorticity generation parameter** for 1 pm CDT on May 31, 2013, the date of the El Reno tornado (marked), showing isobars (black) and VGP (red isopleths and red shading). Even a few hours before the event, VGP patterns had not come together, and that hour-to-hour monitoring may be required in some events.

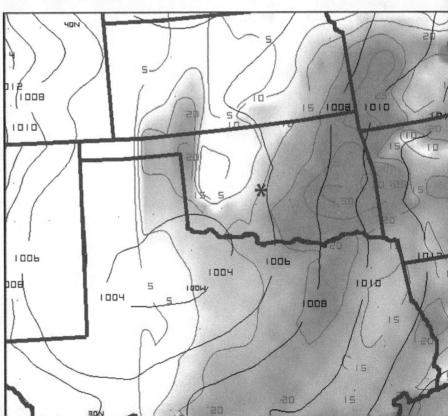

# EHI | Energy-Helicity Index

**Purpose**
- Maximum updraft speed
- Experimental, storm severity

**Abbreviations**
EHI

**Introduced**
John Hart and W. Korotky, 1991

**Maximized**
Very steep lapse rate above the LFC
- "Fat" positive areas

**Units**
m/s$^2$

**Representative values**
| | |
|---|---|
| <1.0 | Supercells/tornadoes unlikely |
| 1.0-2.0 | Supercells/tornadoes possible |
| 2.0-2.4 | Supercells likely |
| 2.5-2.9 | Supercells and tornadoes likely |
| 3.0-3.9 | Strong tornadoes (F2-F3 psbl) |
| >4.0 | Violent tornadoes (F4-F5 psbl) |

**Examples**
| | |
|---|---|
| 0.8 | St Louis hailstorm (4/10/2001) |
| 1.0 | Ohio derecho (6/29/2012) |
| 1.3 | Fort Worth F3 (3/28/2000) |
| 1.4 | Jarrell TX F5 (5/27/1997) |
| 2.5 | Rochelle IL EF4 (4/9/2015) |
| 3.2 | Moore OK EF5 (5/20/2013) |
| 3.2 | Greensburg KS EF5 (5/4/2007) |
| 4.4 | Topeka KS F5 (6/8/1966) |
| 4.9 | Superoutbreak (KY) (4/3/1974) |
| 5.7 | Joplin MO EF5 (5/22/2011) |
| 5.8 | Cen. Oklahoma (5/3/1999) |
| 8.4 | El Reno OK EF3 (5/31/2013) |
| 9.5 | Vilonia AR EF4 (4/27/2014) |

The energy-helicity index, EHI, is a dimensionless number that expresses the synergistic effect of high instability and high storm-relative helicity in producing rotating storms. It was initially implemented in the SHARP workstation (Hart and Korotky 1991). Since EHI is a product of two terms, unusually high values of one term with low values of the other are often associated with significant tornadoes. For example, some tornado events are high-instability low-shear episodes (e.g. the Jarrell, Texas 1997 tornado), and others are high-shear low-instability episodes (e.g., the Illinois tornado outbreak of April 20, 2004 and many cool season nighttime events in the southeast United States).

EHI is defined as follows:

$$EHI = (CAPE \cdot SRH) / 160{,}000$$

where CAPE equals buoyancy in J/kg, and SRH is the storm-relative helicity through a shallow low level layer. The original definition of EHI (1993) specified use of the 0-2 km layer, however it has since been revised by Davies in 2011 to use the 0-1 km layer.

Low values of EHI indicate CAPE and helicity are both low. High values of EHI indicate either CAPE or helicity are high, or that both are high.

Davies 1993 examined a set of tornado cases and determined that values of 2.0 or greater were associated with mesocyclones; 3.0 or greater favored strong tornadoes; and 4.0 or greater violent tornadoes. A more complete list of threshold values is provided in the sidebar.

SEVERE WEATHER INDICES    167

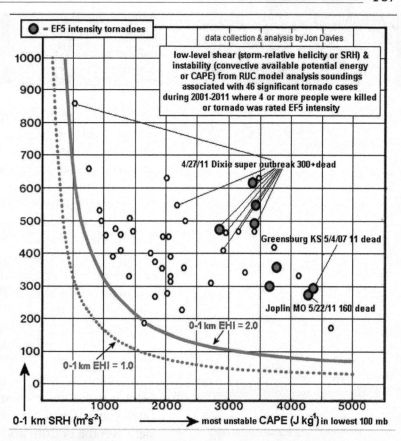

**Figure A1-32 (right): Energy-Helicity Index parameter space** showing CAPE (x-axis) and 0-1 km SRH (y-axis). *(Courtesy Jon Davies)*

**Figure A1-33 (below): Energy-helicity index** for Texas on April 10, 1979, the date of the Wichita Falls F4 tornado (asterisk). Pink dashed contours indicate ML-CAPE, with EHI in thick red, with outer contours in blue and light blue. Surface wind vectors are in light cyan. The storm formed near Seymour, at "x", then tracked toward Wichita Falls. The EHI values show a favorable environment along the length of the dryline.

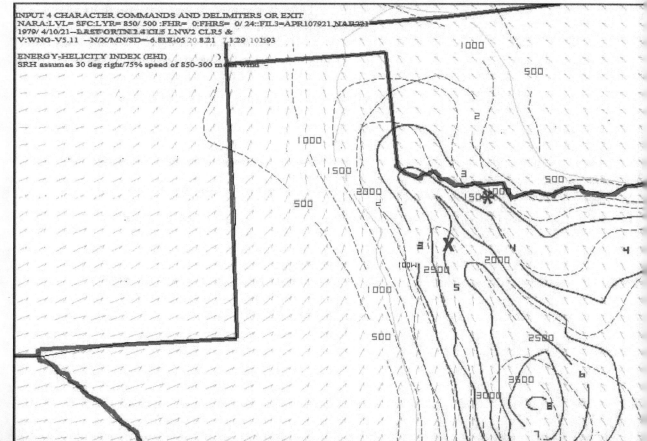

# SCP | Supercell Composite Parameter

**Purpose**
Distinguish non-supercell from supercell environments

**Abbreviations**
SCP

**Introduced**
Richard Thompson; Roger Edwards; and John Hart, 2002

**Maximized**
Very steep lapse rate above the LFC - "Fat" positive areas

**Units**
Dimensionless

**Representative values**
≤0.7 Non-supercells
0.7-1.0 Increasing risk of supercells
5.0 Significant tornadoes likely

**Examples**
0.0 DFW TX / Delta 191 (8/2/1985)
2.8 St Louis hailstorm (4/10/2001)
4.4 Fort Worth F3 (3/28/2000)
5.6 Jarrell TX F5 (5/27/1997)
5.9 Ohio derecho (6/29/2012)
9.2 Greensburg KS EF5 (5/4/2007)
15.4 Joplin MO EF5 (5/22/2011)
18.7 Cen. Oklahoma (5/3/1999)
19.9 Topeka KS F5 (6/8/1966)
20.9 Moore OK EF5 (5/20/2013)
23.8 Rochelle IL EF4 (4/9/2015)
50.5 Superoutbreak (KY) (4/3/1974)
51.0 El Reno OK EF3 (5/31/2013)
78.0 Vilonia AR EF4 (4/27/2014)

The supercell composite parameter (SCP) (Thompson et al 2002; revised in Thompson et al 2004b) was developed to distinguish *non-supercell* from *supercell* environments. As nearly all supercells contain damaging wind and hail, this allows forecasters to rapidly identify areas where severe weather is likely.

The SCP index combines three terms: instability, storm-relative helicity, and bulk shear. It is a relatively simple index that divides each term by a threshold value, so that quantities exceeding the threshold value contribute positively to the SCP, and those below the threshold value reduce the SCP. The SCP is constructed as follows:

$$SCP = (CAPE_{MU}/1000) \cdot (SRH_{EFF}/50) \cdot (BWD_{EFF}/20)$$

where $CAPE_{MU}$ is the most unstable CAPE in J/kg, $SRH_{EFF}$ is the storm-relative helicity using the effective layer method (q.v.) in $m^2 s^{-2}$, and $BWD_{EFF}$ is the bulk shear between the bottom and top of the effective layer. The entire BWD term is set to 0 if BWD is less than 10 m s$^{-1}$, and 1 if it is greater than 20 m s$^{-1}$.

When the SCP value is below 0.7 the environment strongly favors non-supercells. As values rise from 0.7 to 1.0, the risk of supercells increases substantially. At high values, above 5, significant tornadoes are likely.

**Original SCP**
The original Supercell Composite Parameter (Thompson et al 2002) was as follows:

$$SCP = (CAPE_{MU}/1000) \cdot (SRH_{0-3}/150) \cdot (B/40)$$

where $CAPE_{MU}$ is the most unstable CAPE in J/kg; $SRH_{0-3}$ is the storm-relative helicity in the 0-3 km layer; and $B$ is the BRN shear number (0.5 x $U^2$, where $U$ is the shear between 0-0.5 and 5.5-6 km AGL).

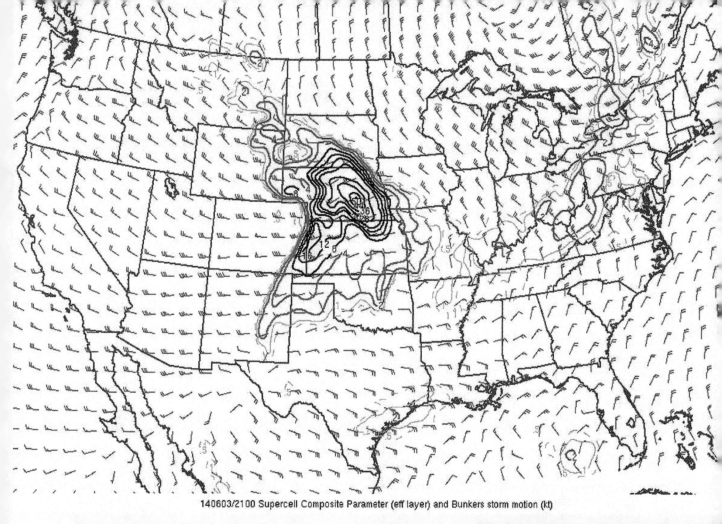

140603/2100 Supercell Composite Parameter (eff layer) and Bunkers storm motion (kt)

**Figure A1-34 (top). Plot of supercell composite parameter** (isopleths) and Bunkers right-mover storm motion (barbs) from SPC analysis of the RAP model output. The situation is for June 3, 2014, when supercells developed in the region northwest of Omaha. In spite of the high SCP, ranging up to 50, only one EF2 tornado was produced It is essential to look at the underlying data to find the components making up elevated SCP values in a target area. SHARPpy provides listings that allow this information to be quickly determined. *(NOAA/SPC)*

**Figure A1-35 (right). Supercell composite** (coloring) and 850 and 500 mb wind plots (barbs). Note that this is a 144-hour GFS forecast plot for May 23, 2015 (24/0000 UTC). Instead of tornadoes near Gage and Fort Stockton, we saw an outbreak of weak tornadoes near Ada, Oklahoma as well as in the Austin, Texas area. No tornaodes were reported in the orange or red areas forecast six days earlier.

# STP | Significant Tornado Parameter

**Purpose**
Distinguish nontornadic from significantly tornadic environments

**Abbreviations**
STP

**Introduced**
Richard Thompson; Roger Edwards; and John Hart, 2002

**Maximized**
- High CAPE
- High bulk shear within the inflow layer
- High SR helicity within the inflow layer
- Low lifted condensation level (high RH)
- Weak capping

**Units**
Dimensionless

**Representative values**
| | |
|---|---|
| ≤1 | Nontornadic storms |
| 1-1.5 | Weakly tornadic supercells |
| 1.5 | Tornadic supercells likely |
| 1.7 | Mean value for sig. tornadoes |

**Examples**
| | |
|---|---|
| -0.6 | Jarrell TX F5 (5/27/1997) |
| 0.0 | DFW TX / Delta 191 (8/2/1985) |
| 0.3 | Ohio derecho (6/29/2012) |
| 0.4 | St Louis hailstorm (4/10/2001) |
| 1.2 | Fort Worth F3 (3/28/2000) |
| 1.3 | Greensburg KS EF5 (5/4/2007) |
| 2.2 | Topeka KS F5 (6/8/1966) |
| 2.4 | Rochelle IL EF4 (4/9/2015) |
| 3.4 | Superoutbreak (KY) (4/3/1974) |
| 4.4 | Moore OK EF5 (5/20/2013) |
| 4.6 | Cen. Oklahoma (5/3/1999) |
| 5.5 | Joplin MO EF5 (5/22/2011) |
| 7.0 | El Reno OK EF3 (5/31/2013) |
| 7.9 | Vilonia AR EF4 (4/27/2014) |

The purpose of the significant tornado parameter, STP, is to distinguish *nontornadic* from *significantly tornadic* environments, particularly in the EF2 to EF5 range. As with SCP it is a multiple-parameter tornado prediction algorithm that considers instability, bulk shear, and storm-relative helicity. However STP adds two additional terms: lifted condensation level and convective inhibition. The STP was introduced in 2002 and revised in 2004 with slightly better results. It is constructed as follows:

$$STP = (CAPE_{ML}/1500) \cdot (BWD_{EFF}/20) \cdot (SRH_{EFF}/150) \cdot ((2000-LCL_{ML})/1500) \cdot ((250+CINH_{ML})/200)$$

where $CAPE_{ML}$ is the mixed-layer CAPE in J/kg; $BWD_{EFF}$ is the effective layer bulk shear (difference between wind between the top and bottom of the layer) in m/s; $SRH_{EFF}$ is the effective layer storm-relative helicity in $m^2/s^2$; $LCL_{ML}$ is the mixed-layer lifted condensation level in meters; and $CINH_{ML}$ is the mixed-layer convective inhibition in J/kg.

There are several conditions: when $LCL_{SB}$ is less than 1000 m, the term is set to 1; when $CINH_{ML}$ is greater than -50 the term is set to 1; when $BWD_{0-6}$ is less than 12.5 the term is set to 0; and the BWD term cannot exceed 1.5.

This is a functionally simple and elegant index which considers five terms, covering instability, lifted condensation level, storm-relative helicity, bulk shear, and convective inhibition. All five of these terms have strong correlation with supercell and tornado risk. Each term is divided by a threshold value, so that values above the value add to the final result, while values below the value diminish the resulting number.

The higher the STP, the greater the risk of significant tornadoes. Values below 1 are associated with nontornadic supercells and weak tornado episodes. As values rise from 1 to 1.5 there is a rapid increase in the risk of significant tornadoes, and above 1.5 tornadoes are almost certain. The mean STP for significant tornado events is 1.7.

# SEVERE WEATHER INDICES   171

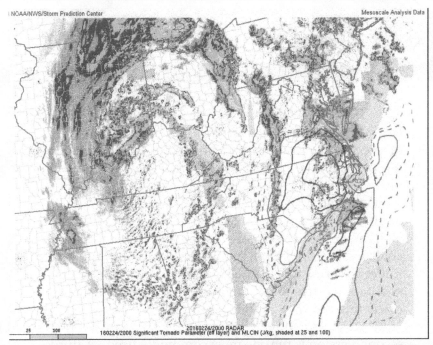

**Figure A1-36. The Storm Prediction Center offers plots of significant tornado parameter (STP)**, as with this example of the February 24, 2016 severe weather event in the Carolinas and Virginia. This shows optimal use of the parameter, with the forecast region being fairly devoid of convective storms and positioned ahead of an approaching line.

**Original STP**
The original Significant Tornado Parameter was as follows, containing slightly different weights and layers (Thompson et al 2002):

$$STP = (CAPE_{ML}/1000) \cdot (BWD_{0-6}/20) \cdot (SRH_{0-1}/100) \cdot ((2000-LCL_{ML})/1500) \cdot ((150-CINH_{ML})/125)$$

where $CAPE_{ML}$ is the mixed-layer CAPE in J/kg; $BWD_{0-6}$ is the 0-6 km bulk shear in m/s; $SRH_{0-1}$ is the storm-relative helicity in the layer 0-1 km; $LCL_{ML}$ is the mixed-layer lifted condensation level; and $CINH_{ML}$ is convective inhibition in J/kg. An LCL less than 1000 m sets the term to 0; a CINH greater than -50 sets the term to 1; and the BWD term has an upper and lower limit of 1.5 and 0, respectively (i.e. BWD velocity of 30 and 12.5 m/s).

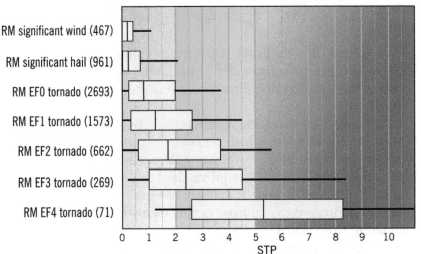

**Figure A1-37. Verification of the significant tornado parameter.** This shows its effectiveness in predicting strong tornadoes, with virtually no overlap between the mid-spread of the EF4 category and that of the EF1 and weaker categories.

**SCP versus STP quick summary**

**Supercell composite param. (SCP)** distinguishes supercell from non-supercell environments. The threshold value between these two categories is 0.7.

**Significant tornado param. (STP)** distinguishes significant tornado environments from nontornadic environments. The threshold value is 0.5 to 1.0.

# SWEAT | Severe Weather Threat Index

**Purpose**
- Maximum updraft speed
- Experimental, storm severity

**Abbreviations**
SWEAT

**Introduced**
V1, Miller and Bidner, 1970
V2, Miller end Bidner, 1971
V3, Miller and Robert Maddox, 1975

**Maximized**
Very steep lapse rate above the LFC
- "Fat" positive areas

**Versions**
V1, 1970. No shear term.
V2, 1972. Adds the shear term.
V3, 1975. High terrain version.

**Units**
m/s$^2$

**Representative values**
*from Miller et al 1970*
0-299    Severe storms not expected
300-399  Severe thunderstorms
400+     Tornadoes

**Examples**
139   Jarrell TX F5 (5/27/1997)
232   DFW TX / Delta 191 (8/2/1985)
387   St Louis hailstorm (4/10/2001)
405   Rochelle IL EF4 (4/9/2015)
438   Vilonia AR EF4 (4/27/2014)
458   Topeka KS F5 (6/8/1966)
477   Moore OK EF5 (5/20/2013)
501   Ohio derecho (6/29/2012)
530   Superoutbreak (KY) (4/3/1974)
543   Fort Worth F3 (3/28/2000)
546   Joplin MO EF5 (5/22/2011)
556   Cen. Oklahoma (5/3/1999)
595   El Reno OK EF3 (5/31/2013)
703   Greensburg KS EF5 (5/4/2007)

The SWEAT Index was developed by the U.S. Air Force Military Weather Warning Center for forecasting severe weather in the central United States. It represents one of the first attempts to incorporate shear into an index. At the time it was developed, all indices in common use were thermodynamic. It is based on a study of 328 tornado events, and as null events were not factored in, it only confirms the existence of a favorable tornado environment. In the original document on SWEAT index, Miller confirms this, saying, "nothing can be known about false alarm rates" and warned against its use for predicting ordinary thunderstorms.

There are three versions of the SWEAT index. The second version is the one most commonly used. A third version was developed (Miller and Maddox 1975) which replaced the properties at 850 mb with those at 900 m AGL, allowing for use in higher terrain, with alterations in the shear term. It is not in use today.

The second version, which constitutes the customary definition of SWEAT, is as follows:

$$\text{SWEAT} = 12T_{d850} + 20(TT-49) + 2f_{850} + f_{500} + 125(S+0.2)$$

where TT is the Total Totals Index value, $f$ is the ground-relative wind speed in knots, and S equals the sine of the 500 mb wind direction minus the 850 mb wind direction.

SWEAT uses a crude wind vector between the 500 mb and 850 mb wind, which is above the critical tornado inflow layer but still captures some elements helicity. The shear term $125(S+02)$ requires all of the conditions in Table XX-X to be met. If they are not met, the shear term is set to zero to eliminate erratic contributions to the equation from unusual patterns.

Data gathered from 102 storm events showed that SWEAT values of 300 or higher favored severe thunderstorms, and 400 or higher indicated tornadoes. Independent verification showed values about 20% lower were applicable to northeast United States thunderstorm cases.

SEVERE WEATHER INDICES  173

**SWEAT index shear term requirements**

• the 850 mb wind direction must equal 130-250°
• the 500 mb wind direction must equal 210-310°
• both the 850 mb and 500 mb wind speeds must equal ≥ 15 kt
• the term $f_{500} - f_{850}$ mb must be greater than zero (winds veering with height)

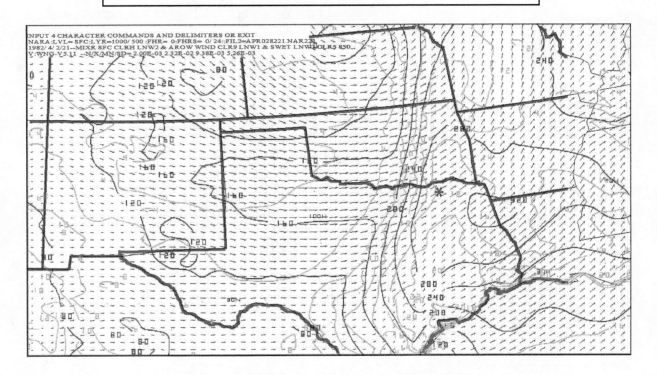

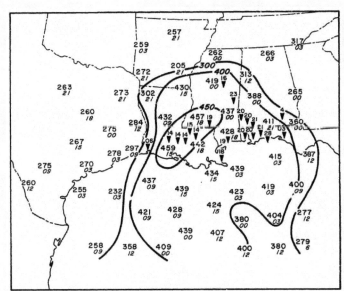

**Figure A1-38 (above). SWEAT index** on April 2, 1982, at 2100 UTC, when Paris, Texas (asterisk) was struck by an F4 tornado. SWEAT is in dark red, with surface mixing ratio in green and surface wind vectors in cyan. The boundary from Topeka to Dallas is a dryline, with a cold front close behind. The SWEAT values of 260 were marginal, emphasizing the need to use other parameters as well as to analyze the constituent shear and helicity fields.

**Figure A1-39. SWEAT index** verification for September 16, 1971. A total of 17 tornadoes were recorded, with one F3 and two F2 tornadoes in Louisiana, and two rated F2 in Alabama. The adaptability of the SWEAT index to numerical model output made it highly useful during the 1970s and 1980s. (From Miller 1975)

# SHIP | Significant Hail Parameter

**Purpose**
- xxx

**Abbreviations**
SHIP

**Introduced**
xxx

**Maximized**
xxx

**Units**
xxx

**Representative values**
>1.0  Favorable for significant hail
1.5  Significant hail (>2.0")
2.0  Significant hail (>2.0")
      tennis ball possible

**Examples**
0.0  DFW TX / Delta 191 (8/2/1985)
0.6  St Louis hailstorm (4/10/2001)
1.2  Topeka KS F5 (6/8/1966)
1.4  Rochelle IL EF4 (4/9/2015)
1.7  Superoutbreak (KY) (4/3/1974)
1.7  Vilonia AR EF4 (4/27/2014)
2.0  El Reno OK EF3 (5/31/2013)
2.1  Cen. Oklahoma (5/3/1999)
2.2  Fort Worth F3 (3/28/2000)
3.3  Jarrell TX F5 (5/27/1997)
3.4  Ohio derecho (6/29/2012)
3.7  Moore OK EF5 (5/20/2013)
3.8  Greensburg KS EF5 (5/4/2007)
3.8  Joplin MO EF5 (5/22/2011)

The significant hail parameter (SHIP) forecasts the likelihood of 2-inch or larger hail (Blumburg et al 2017). It is based on a very large database of hail reports, and uses instability, water vapor, lapse rate, mid-level temperature, and shear to forecast hail environments.

*Although SHIP produces values that look like sizes, it does not forecast hailstone size.* Values of 1.0 indicate a favorable general hail environment; 1.5 to 2.0 indicates significant hail potential; and 4.0 indicates very high potential.

The parameter is quite new so very little information on the evaluation and performance of SHIP was available at press time, however its greatest strength appears to be when used with mesoscale model fields, such as the RAP fields that make up the SPC Experimental Mesoanalysis pages.

**Definition**

The equation is defined as follows:

$$\text{SHIP} = \frac{\text{CAPE}_{MU} \cdot q_{MU} \cdot \Gamma_{500\text{-}700} \cdot T_{500} \cdot S_{0\text{-}6}}{R}$$

$$44{,}000{,}000$$

where $\text{CAPE}_{MU}$ is the most unstable CAPE; $q_{MU}$ is the mixing ratio at the most unstable level (j/kg); $\Gamma_{500\text{-}700}$ is the lapse rate between 700 and 500 mb; $T_{500}$ is the 500 mb temperature; $S_{0\text{-}6}$ is the shear magnitude between 0 and 6 km in m/s; and $R$ equals a constant, 44,000,000.

The index has a number of constraints: if the term $S_{0\text{-}6}$ exceeds 27 m s$^{-1}$ it is set to that value; likewise if it is lower than 7 m s$^{-1}$, it is set to that value. Also $q_{MU}$ cannot be lower or higher than 11.0 or 13.6 g kg$^{-1}$, and if so it must be set to the lower or upper value as appropriate. Finally if $T_{500}$ is warmer than -5.5 °C, it is set to that value.

Finally, the definition specifies a set of "secondary modifications". If $\text{CAPE}_{MU}$ is lower than 1300 J kg-1, SHIP is set to SHIP * ($\text{CAPE}_{MU}$/ 1300). If $\Gamma_{500\text{-}700}$ is lower than 5.8 C° km-1 then SHIP is set to SHIP * ($\Gamma_{500\text{-}700}$/ 5.8). And finally if the freezing level $z$ is less than 2400 m AGL, then SHIP is set to SHIP*(z/2400).

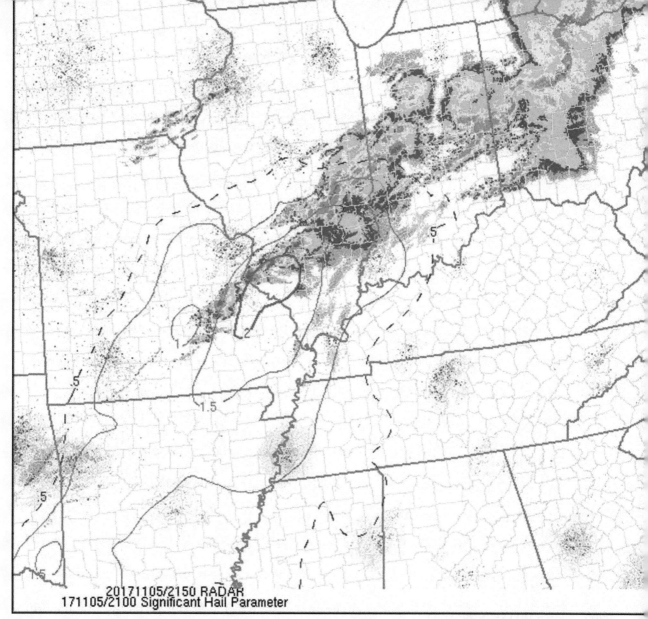

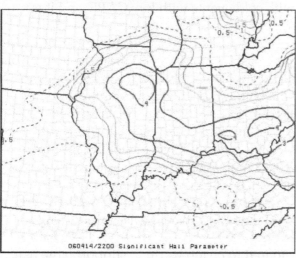

**Figure A1-40 (above): Significant hail parameter** plot for November 5, 2017, showing severe thunderstorm cells, with 2-inch SHIP probabilities indicating a good likelihood of 2-inch hail.

**Figure A1-41 (left): Significant hail parameter** for April 14, 2006. Hail between 2.5 to 3 inches was reported in east central Illinois into northwest Indiana along a west-northwest to east-southeast warm front. The SHIP values of 3 to 4 are consistent with a forecast for large hail.

# WBZ | Wet Bulb Zero height

**Purpose**
- xxx

**Abbreviations**
WBZ

**Introduced**
Fawbush and Miller, 1953

**Maximized**
Low mid-level temperatures
Low mid-level relative humidity

**Units**
xxx

**Representative values**
<6000 ft    Small hail
7000 ft     Large hail (lower threshold)
8000 ft     Large hail potential
9000 ft     Large hail potential
10,000 ft   Large hail (upper threshold)
11,000 ft   Diminished hail size
12,000 ft   Small hail
>12,000 ft  Hail unlikely; WBZ too high

**Examples** (ft AGL)
8475    Fort Worth F3 (3/28/2000)
8600    Cen. Oklahoma (5/3/1999)
9000    Rochelle IL EF4 (4/9/2015)
9200    Joplin MO EF5 (5/22/2011)
9200    Greens. KS EF5 (2007-05-04)
9300    St Louis hailstorm (4/10/2001)
10000   Moore OK EF5 (5/20/2013)
10700   Topeka KS F5 (6/8/1966)
11000   Superoutbreak (KY) (4/3/1974)
11430   Jarrell TX F5 (5/27/1997)
11500   Vilonia AR EF4 (4/27/2014)
11600   El Reno OK EF3 (5/31/2013)
12200   Ohio derecho (6/29/2012)
13400   DFW / Delta 191 (8/2/1985)

The wet-bulb zero height identifies the height above ground level (AGL) at which the environmental temperature, $T_w$, equals 0°C. It is assumed that hailstones below this level do not melt, as even if the temperature is above freezing, evaporation would absorb latent heat, chilling the hailstone. However if $T_w$ is above 0°, hailstones will shed mass in the form of water; that is, they will melt as they fall. The ideal range for large hail is 7,000 to 10,000 ft AGL, with the range in general for hail 6,000 to 12,000 ft AGL.

Wet bulb zero height is loosely related to the freezing level. However the freezing level strictly considers temperature, while wet-bulb temperature is a combination of temperature and dewpoint. If the temperature is 37°F and dewpoint is 4°F, the temperature would suggest melting, however the dry conditions would result in strong evaporative cooling, and hail would remain in the solid state.

In a typical Great Plains and Midwest air mass, a very high WBZ, above 12,000 ft, indicates the hailstone will fall through a very deep column over a long period of time, and very little of the hailstone will remain, or what was hail will change in the form of rain or a cold downdraft before it reaches the ground.

On the other hand, low WBZs, below 6,000 ft, are associated with cold events with low CAPE, and it is unlikely, though not improbable, that large hail will occur.

Miller 1972 identified an important exception with tropical "Gulf Coast" air masses; 89% of the events in this region was associated with the 11,000 to 14,000 ft layer. Hail sizes were below 1/2". It is important to note that the Gulf Coast air masses are not necessarily in the Gulf Coast, but rather have "tropical" characteristics with very deep moisture and modest, moist adiabatic lapse rates.

The 8,500 ft AGL WBZ line, when plotted on an analysis chart, can be considered to be a useful axis for locating storms with maximum hail size. Furthermore, Miller specified the 10,500 ft isoline as a useful "cut-off" line for hailstorms, Equatorward of this isoline, the air mass is considered to be too warm to support large hail.

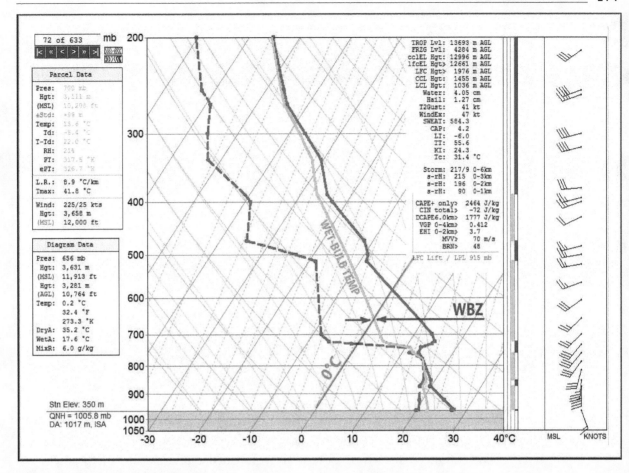

**Figure A1-42 (above): Skew-T log-p diagram** for June 22, 2003 at Omaha, Nebraska, at the time that the largest hailstone at the time fell from a supercell in eastern Nebraska. The hailstone measured 7 inches in diameter, breaking the record set at Coffeyville, Kansas in 1970. The wet bulb zero profile is plotted in green between the T and $T_d$ lines. The wet bulb zero height measures 11,900 ft, placing it well outside the expected limits and serving as a reminder there always will be exceptions.

**Figure A1-43 (right): Wet bulb zero study** from Miller 1972, based on an extensive list of storm events selected by Col. Miller. The 7000-9000 ft level was identified as most associated with large hail.

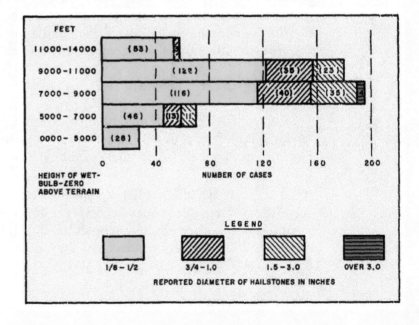

# Additional indices

Provided here are newer indices that are not common or have not gained widespread acceptance. However many of them appear in various journals or are implemented in tools such as SHARPpy or RAOB.

### A1.3.1. Wind damage parameter (WNDG)

The wind damage parameter (source unknown, probably an internal SPC parameter) examines looks for high instability, strong lapse rate, enhanced mid-level flow, and weak inhibition. It is constructed as follows:

$$\text{WNDG} = (\text{CAPE}_{ML} / 2000) * (\Gamma_{0-3} / 9) * (V_{1-3.5}/15) * ((50 + \text{CIN}_{ML})/40)$$

where $\text{CAPE}_{ML}$ is the 100 mb mixed-layer CAPE in J kg$^{-1}$; $\Gamma_{0-3}$ is the 0-3 km lapse rate in °C or K; $V_{1-3.5}$ is the mean 1000-3500 m wind in m s$^{-1}$; and $\text{CIN}_{ML}$ is the 100 mb mixed-layer CIN in J kg$^{-1}$. If the lapse rate $\Gamma_{0-3}$ is less than 7 °C then the term is set to 0; and $\text{CIN}_{ML}$ is set to 50 J kg$^{-1}$ if the value is less than 50.

When WNDG exceeds 1, then there is an enhanced risk for scattered damaging outflow gusts with multicell clusters.

### A1.3.2. Enhanced stretching potential (ESP)

The enhanced stretching potential (from Caruso and Davies 2005) determines where buoyancy and steep lapse rates exist. This set of ingredients is thought to be conducive to low-level vortex stretching and possible tornadoes from *non-mesocyclone processes* (possibly including landspouts) as well as standard mesocyclone tornadoes. High values along a slow-moving or stationary boundary are especially favorable for vortex development. The equation is:

$$\text{ESP} = (\text{CAPE}_{ML} / 50) * ((\Gamma_{0-3} - 7) / 1)$$

where $\text{CAPE}_{ML}$ is the 100 mb mixed-layer CAPE in J kg$^{-1}$; and $\Gamma_{0-3}$ is the 0-3 km lapse rate in °C or K. ESP is set to zero whenever the CAPE is less than 250 J kg$^{-1}$ or the lapse rate is less than 7 C° km$^{-1}$.

### A1.3.3. Non-supercell tornado parameter

The non-supercell tornado (NST) parameter (Baumgardt and Cook 2006) is similar to enhanced stretching potential in that it attempts to identify environments conducive to non-mesocyclone tornadoes. It was designed to build on ESP, and was written to be used in AWIPS. The index is constructed as:

$$NST = (\Gamma_{0-1} / 9) * (3CAPE_{ML} / 100) * ((225 - CIN_{ML}) / 200) * ((18 - S_{0-6}) / 5) * (\zeta_r / 8)$$

where equals the 0-1 km lapse rate (°C km$^{-1}$); $3CAPE_{ML}$ equals the CAPE lifted to 3 km for a 0-1 km mixed parcel (J kg$^{-1}$); $CIN_{ML}$ equals the convective inhibition for a 0-1 km mixed-layer parcel (J kg$^{-1}$); $S_{0-6}$ equals the bulk shear in the 0-6 km layer (m s$^{-1}$); and $\zeta_r$ equals the surface relative vorticity ($10^{-5}$ s$^{-1}$).

One of the benefits of NST is the use of the vertical vorticity parameter, which forces the equation to focus on boundaries and mesolows, while minimizing contributions away from boundaries and mesolows. However it is sensitive to the analysis algorithm, which may produce differing vorticity results depending on which one is selected.

Values of NST above 1 are favorable for vortices within thunderstorms. Values above 2 suggest enhanced potential. However the exact ranges have not been fully evaluated.

### A1.3.4. MCS maintenance probability (MMP)

The MCS maintenance probability parameter (Coniglio 2007) identifies the possibility for MCSs to maintain peak intensity for at least an hour.

The equation is more complex than most other severe weather equations, and it cannot easily be calculated by hand due to the use of terms like maximum shear, but it is defined here so readers can examine the contributing quantities. It is constructed as follows:

$$MMP = 1 / (1 + \exp\{a_0 + [a_1(S_{MAX})] + [a_2(\Gamma_{3-8})] + [a_3(CAPE_R)] + [a_4(V_{3-12})]\})$$

where $a_0 = 13.0$; $a_1 = -0.0459$; $a_2 = -1.16$; $a_3 = -0.000617$; $a_4 = -0.17$; $S_{MAX}$ equals the maximum deep-layer shear (s$^{-1}$); $\Gamma_{3-8}$ equals the 3-8 km lapse rate (°C km$^{-1}$); CAPE equals the best CAPE for the environment (J kg$^{-1}$); and V3-12 equals the mean 3-12 km wind speed (m s$^{-1}$).

### A1.3.5. SigSvr

The significant severe, or SigSvr parameter (Craven and Brooks 2004) attempts to combine instability with 0-6 km shear, in a similar manner to EHI and helicity. However since this evaluates shear instead of helicity, it is better for finding general severe weather potential. The equation is:

$$SIGSVR = CAPE_{ML} * S_{0-6}$$

where $CAPE_{ML}$ is the 100 mb mixed layer CAPE, and $S_{0-6}$ is the bulk shear between the top and bottom of the 0-6 km layer.

Values of above 10,000 are favorable for severe thunderstorms; above 20,000 indicate high winds and hail; and above 30,000 suggest significant tornadoes.

# Obsolete indices

Many indices were developed during the early years of storm forecasting, and most of slipped into obscurity. Some of these are summarized in Peppler (1988), an excellent reference. The more notable indices are as follows:

## A1.4.1. Rackliff Index

The Rackliff Index (Rackliff 1962) was the first British stability index tailored for forecasting summertime thunderstorms, and was based on a dataset of 89 days of 2300 UTC radiosondes launched in 1959. It uses the wet bulb potential temperature at 900 mb, the level having been selected for being above the nighttime radiational inversion.

$$RACK = \theta_{w900} - T_{500}$$

Following are the values determined by the initial study. It was found that heavy thunderstorms occurred within the range of 31 to 35, suggesting 30 as a transition value.

&lt;25  No showers
26-29 &lt;50% chance of showers
30-31 &gt;50% chance of showers
&gt;32  100% chance of showers

## A1.4.2. Jefferson Index

The Jefferson Index (Jefferson 1963a) is a modification of the Rackliff Index developed for British forecasting. It was constructed so as to give the same threshold values for thunderstorms over a wide range of temperatures. The format has some similarities to the K Index.

Values of 28 to 29 were found to be appropriate as a threshold value for the presence of thunderstorms.

$$JEFF = 1.6 * \theta_{w900} - T_{500} - 0.5(T_{700} - T_{d700}) - 8$$

## A1.4.3. Adedokun Index

The Adedokun Index (Adedokun 1981, 1982) was developed for forecasting precipitation in west Africa.

$$ADED1 = \theta_{w850} - \theta_{s500}$$
$$ADED2 = \theta_{wSfc} - \theta_{s500}$$

The value $\theta_s$ is obtained by lowering a 500 mb parcel moist adiabatically to 1000 mb.

The ADED1 Index was found to be better for identifying non-occurrence of precipitation, while ADED2 was better for forecasting precipitation.

## A1.4.4. Total energy index

The total energy index was introduced in Darkow 1968. It is computed as:

$$ET = (E_{T500} - E_{T850})$$

The energy of the air is obtained in cal g$^{-1}$. This of course needs specialized software to compute the values.

Typical total energy values are:
0 to -1  Thunderstorms possible
-1 to -2  Isolated severe
&lt;-2  Severe storms; tornado activity

## A1.4.5. Lid strength index

The Lid Strength Index was devised by (Carlson et al 1980). It is as follows:

$$\text{LSI} = (\theta_{wa} - \theta_{sw5}) - (\theta_{swl} - \theta_w)$$

where $\theta_w$ is the WBPT (wet bulb potential temperature, i.e. moist adiabat value) at the surface, $\theta_{wa}$ is the averaged WBPT in the lowest 50 mb of the atmosphere, $\theta_{sw5}$ is the WBPT at 500 mb, $\theta_{swl}$ is the WBPT at the warmest point of the inversion between 850 and 500 mb.

This reflects one potential source of problems: three of the four terms use fixed upper level layers, so abnormal lid heights will result in erroneous or unrepresentative values. The index also does not take into account elevated situations.

A LSI of below -2.0 is correlated with a lack of inhibition, while above +2.0 indicates capping will suppress storms. This value of +2 corresponds with the 90th percentile value found by Carlson.

The LSI was quickly replaced by CIN during by the 1990s as computing power necessary to do the integrations with depth was becoming widely available and was immune to the fixed layer issues presented by LSI.

## A1.4.6. Fawbush-Miller index

The Fawbush-Miller Index for storm forecasting

>0.4    No precipitation
0.4...-2.4    Showers
-2.5...-6.4    Thunderstorms
<-6.4    Severe thunderstorms

## A1.4.7. Fawbush & Miller hail method

The Fawbush-Miller method for hail (Fawbush and Miller 1953) is included because of its widespread use over many decades for estimating hail size. It still has value as a quick reference technique to obtain information for a hail forecast. It considers characteristics of the temperature profile above the convective condensation level (CCL) and relates this to the -5°C altitude. The technique is described in depth in Figure A1-44.

The main limiting factor with this method is the proximity souding problem: computations made with the standard 0000 and 1200 UTC soundings are often not representative of the storm environment later in the day, and may not be representative of the environment at the storm location.

Since sounding modification is required, the method is not suitable for web sites that make automatic computation on soundings. However, useful forecasts can be made from model grids. Forecasters working up a mesoscale forecast should perform the usual modifications to soundings to reflect conditions at peak heating before using the Fawbush Miller hail method.

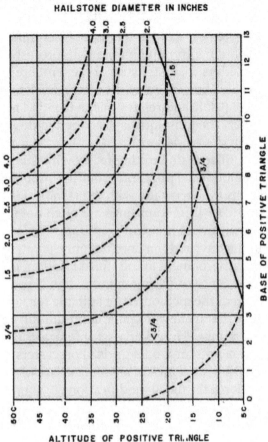

Figure 48 The Fawbush-Miller Hail Graph showing the forecast hailstone diameter in inches. Graph revised November 1965 on the basis of 622 hail reports.

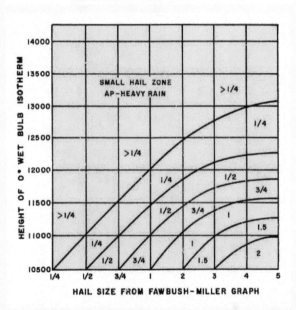

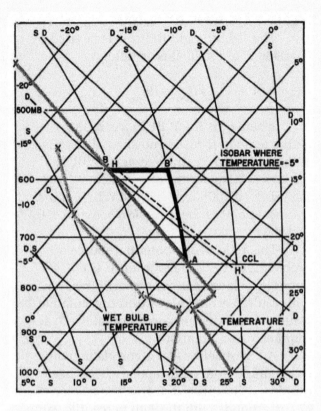

**Figure A1-44. Fawbush-Miller hail size method.**

**Triangle construction (above).** The diagram (above) above shows how the triangle is constructed. The lowest vertex, A, is constructed at the CCL level on the temperature line. The triangle side A-B' is drawn upward parallel to the moist adiabats. The top left vertex, B, is drawn where the temperature profile equals -5°C. The side B-B' is drawn from this vertex along an isobar to the moist adiabat. Where they intersect forms the vertex B'. The temperature at A, B, and B' are read from the diagram and written down. Subtract B from B' to obtain the "base value", and subtract B from A to obtain the "altitude value", then proceed to the computation.

**Computation table (top left).** The characteristics of the triangle are consulted on the table at the top left. The intersection of the altitude and base value are located and read against the curved lines, which give hail size.

**Tropical air mass table (left).** The square table at left with heights from 10,500 to 14,000 ft are used strictly for Miller's Type II air masses, or "Gulf Coast" or tropical air masses. This should be used whenever the wet-bulb zero height is above 10,500 ft.

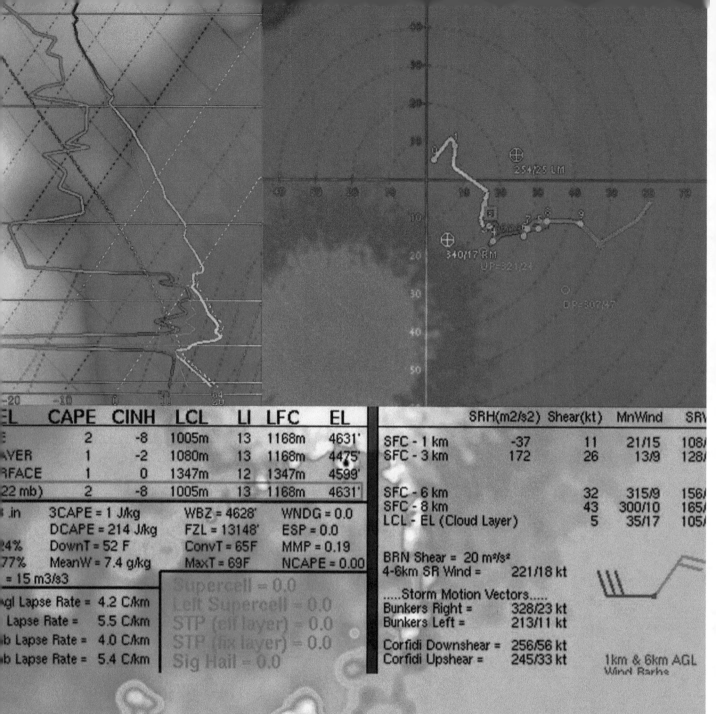

# A-2 SHARPPY

The SHARPpy, Sounding and Hodograph Analysis and Research Program in Python (Blumberg, 2017), is a revolutionary standardized display suite of Skew-T, hodograph, and index data that entered common use during the mid-2010s on a variety of government and private sector Internet web sites. It is an open source library written in the Python programming language, which marks a departure from the FORTRAN and C++ libraries that have dominated NOAA code libraries for decades.

Developing SHARPpy as an application would have severely limited its usability, locking it to a specific platform and user input method. Instead, it was developed as a framework. This means that the code is reusable and other scientists, programmers, and researchers can build applications and scripts around it, obtaining the very same numerical results regardless of who writes the applications.

Furthermore, SHARPpy is an outgrowth of the SHARP program which was developed within NOAA in the 1990s and ran under MS-DOS. Upgrading it further would leave little room for extensibility, scalability, and portability to the Web and other operating systems.

As SHARPpy has been widely accepted in forecast systems on the Internet and in applications using the code library, it is essential for forecasters to have a thorough understanding of every component of the SHARPpy display. This is what we will cover in this chapter.

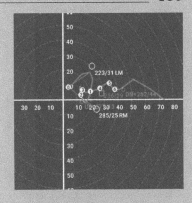

**SHARPpy authors**
SHARPpy is a Python framework initially developed in 2011 and has been improved since then.

**Original developer**
Patrick Marsh / SPC

**Current co-developers**
Kelton Halbert / OU SOM
Greg Blumberg / OU SOM

**Testing and debugging**
David John Gagne / NCAR
Timothy Supinie / OU

**Datasets and other contributors**
Richard Thompson / SPC
Ryan Jewell / SPC
Bryan Smith / SPC
Matthew Bunkers / NWS
John Hart / SPC

**Program location**
github.com/sharppy/SHARPpy

## A2.1. SHARPpy

During the 1980s and 1990s, National Weather Service forecasters enjoyed the unprecedented ability to plot skew-T's on AFOS (Automation of Field Operations and Services) workstations. This provided a basic set of stability indices. However as the code base was centrally maintained, this limited the ability of forecasters to take advantage of new severe weather techniques, particularly when dealing with shear and helicity.

PC computers began appearing in National Weather Service offices in the mid-1980s, which opened possibilities for developing programs and sharing them between offices. One of the first sounding viewers to be developed was SHARP, developed by NOAA forecasters John Hart and William Korotky in 1991. This program was widely used for

> **SHARPpy color codes**
> This table should be used as a quick referene to the hodograph, storm slinky, and other lines plotted on the chart.
>
> | | |
> |---|---|
> | Red | 0-3 km AGL |
> | Green | 3-6 km AGL |
> | Yellow | 6-9 km AGL |
> | Lt. Blue | 9-15 km AGL |
> | Purple | Equilibrium Level |

a decade, and its output appeard in many journal articles. However its life cycle began coming to an end in the early 2000s with the decline of MS-DOS applications.

Although some standardization of sounding and hodograph analysis was achieved on AWIPS (Advanced Weather Interactive Processing System, the primary mainframe tool for National Weather Service forecasters), this was cumbersome and only offered limited accessibility. Outside of NOAA, thermodynamic and sounding libraries were in many different formats. Some were commercial proprietary tools, others were open source but were not adaptable for Web development, and others were developed in foreign countries where language barriers introduced the potential for errors.

More worrisome, it was also found that when tested, the results from many of the available packages yielded different values for the very same station, date, and configuration. These kinds of inconsistencies degrade the ability of the meteorological community to conduct proper analysis and research.

The need for a solution to this predicament was SHARPpy. Work on it began in 2011 and a proof-of-concept version was shown in 2012 at the American Meteorological Society annual meeting. An application limits users to work with the data as the designers intended, but a framework allows users to structure their data and applications in almost any way conceivable, while obtaining accurate calculations of thermodynamic and shear values and other severe weather parameters.

## A2.2. Concept of operation

Input to SHARPpy is provided as the core elements of a thermodynamic and shear profile: temperature, dewpoint, pressure, height, wind speed, and wind direction data (Blumberg et al, 2017). This eliminates the dependencies on decoding various data formats and allows the library to be used with a wider variety of data such as AMDAR, dropsondes, model output; the user only needs to write out the elements.

However SHARPpy does draw on other sources of data. User input may be needed, such as to choose from different parcel lifting methods or to determine the storm vector. The framework may also refer back internally to his-

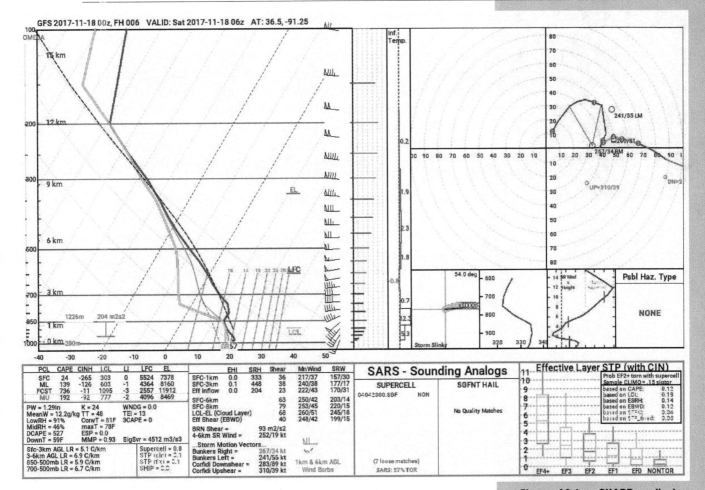

**Figure A2-1. SHARPpy display**, showing all of the frames. This is the Pivotal Weather configuration, which has minor differences between other SHARPpy displays. This is also the output from a model field, rather than a radiosonde observation. *(Pivotal-Weather.com)*

torical data, such as the Sounding Analog Retrieval System, which as we shall see is an analog forecasting technique.

In the following sections, note that the ordering may vary slightly between various SHARPpy implementations depending on how it is configured. The basic concepts however still fully describe the products.

## A2.3 Sounding

SHARPpy presents a conventional Skew T log P diagram at the top left of the image. However in addition to wind plots and height labels, it includes scales showing vertical motion, shear, temperature advection, and other parameters. Being familiar with all of these quantities, scales, and indicators will help the forecaster fully exploit the data.

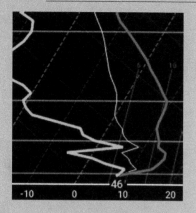

### A2.3.1. Pressure levels

In addition to the standard pressure levels, several key levels are marked. These include the lifted condensation level (green), the level of free convection (yellow), and equilibrium level (purple). The effective inflow layer, if present, is outlined in cyan and this is duplicated on the hodograph. A color code height scale on the right side relates colors for this sounding to height above ground level (AGL). A scale on the left side printed in red actually displays the AGL heights.

### A2.3.2. Temperature and dewpoint plots

The plotted data in the diagram itself contains temperature profile, in red, and a dewpoint profile, in green. There is also a wet bulb temperature profile, which appears as a thin cyan line between the temperature and dewpoint. This can be useful for manually working up DCAPE calculations or using hail prediction algorithms.

### A2.3.3. Omega profile

If the input data source supplies vertical motion, omega (Pa s$^{-1}$), SHARPpy will display the information on a purple scale along a vertical axis marked "OMEGA" at the far left. At each level, a bar will extend horizontally. If it is red and points to the right, omega is negative (pressure decreasing per unit of time) indicating rising motion at that level. If it is blue and points to the left, omega is positive, indicating subsidence at that level; sinking motion. The magnitude of the bar is proportional to the magnitude of vertical motion, and if no bar is seen, omega is close to zero. It is important to keep in mind that these motions are approximations; the terms used to calculate omega are where the model has the most difficulty, and they tend to cancel one another out.

### A2.3.4. Wind speed profile

The wind speed scale is positioned immediately to the right of the Skew T diagram along a vertical axis, and is marked with color-coded (Table X-XX) wind speeds. These are ground-relative wind speeds, in other words, exactly as reported in radiosonde and flight wind data. The further the bar extends to the right, the stronger the ground-relative wind speed. A scale of dashed brown vertical lines is calibrated every 20 knots. The absence of a bar indicates calm winds or no data.

### A2.3.5. Geostrophic temperature advection profile

Further to the right of the wind speed profile is another vertical axis made up of red and blue segments and labeled at the top "Inf. temp" (inferred temperature advection). Values are in C° h$^{-1}$. Red segments indicate warm air advection and blue segments indicate cold air advection. The values are calculated based on veering or backing of the layer, and the sounding is assumed to be geostrophic balance.

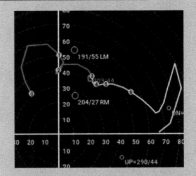

### A2.3.6. Parcel ascent

The most unstable (MU) parcel virtual temperature ascent line (white dashed), downdraft parcel virtual temperature descent line (purple dashed).

### A2.3.7. Virtual temperature

One significant benefit of SHARPpy is that it uses virtual temperature in all CAPE and CIN calculations.

## A2.4. Hodograph

A conventional hodograph is plotted in the top right of the SHARPpy diagram. The origin (0,0) of this diagram is the ground. The hodograph plot is color coded (Table XX-X) according to the AGL height. Dots will be plotted along the line, allowing forecasters to identify the altitude of each hodograph segment in km AGL.

A brown square indicates the mean wind between the LCL and the EL. This is simply an average between the top and bottom of the layer.

Two white circles will appear, one marked RM (right motion) and the other LM (left motion). These indicate the Bunkers method storm motion (Bunkers et al 2014) for deviant right and left moving storms. The 0-6 km mass-weighted storm motion used in the Bunkers calculation is not plotted but can be assumed to be midway between the RM and LM vectors.

Two blue marks will be plotted: DN (downshear) and UP (upshear), representing the Corfidi downshear and upshear vectors (Corfidi 2003). These estimates of mesoscale convective system (MCS) movement and are described in previous chapters.

If an effective inflow layer (Thompson et al 2007) is identified, it will show in cyan as a pie slice, drawn from the storm motion vector to the bottom and top layers of the hodograph that form the effective inflow layer.

Also in certain situations a purple line will appear on the hodograph, representing the 0-500 m AGL shear vector (Esterheld and Giuliano 2008). If this is depicted, the critical angle parameter will also be printed in light blue at the bottom left. Both of these parameters will appear whenever the effective inflow layer is surface-based.

## A2.5. Storm slinky

The Storm Slinky is a technique for plotting parcel displacement in the horizontal plane during its upward ascent in the updraft. When viewed as a whole, it provides a 3-D trajectory of the updraft. It is plotted as a series of circles, and is typically at the top right of the SHARPpy plot although not all implementations will carry it.

The Storm Slinky calculation technique (Blumberg, 2017) begins at the level of free convection (LFC), starting with a 5 m s$^{-1}$ (9.7 kt) upward motion. Buoyant ascent of the parcel is allowed to occur, and ends when it reaches its equilibrium level (EL). During the ascent, the storm-relative winds are allowed to act on the parcel and displace it in two-dimensional space. For example, if the radiosonde data observes a southeast wind at all levels, the parcel will take on an increasingly northwest trajectory away from the storm slinky center.

When this is completed, updraft tilt is then calculated by measuring the x,y,z location of the parcel at the LFC and the x,y,z location at the EL, and determining the tilt with respect to the ground (the larger the tilt number, the closer to 90° and the more erect the updraft).

A white line appears in the display, displaying the Bunkers right-movement storm motion vector. It points to the direction of motion, so when it points southwest, storms can be expected to propogate from northeast to southwest. The magnitude indicated by this line appears constant on the diagram but is proportional to storm motion.

In the SHARPpy Storm Slinky display, coloring is coded by absolute height (Table XX-X), with warm shades (red and green) assigned to the lower troposphere, green shades

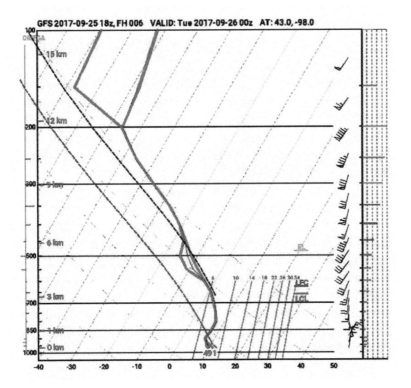

**Figure A2-2. Skew-T diagram from SHARPpy.** Note that this shows a frontal inversion, with a strong stable layer between 900 and 800 mb, and with the dewpoint line closely following the temperature line.

to the middle troposphere, and cold shades (blue and purple) assigned to the upper level. Therefore, if the storm slinky lacks any red color, this is an immediate indication that the parcel is elevated. If the entire column is negatively buoyant, then no storm slinky will appear.

The shape of the Storm Slinky yields important information about the removal of the precipitation cascade from the updraft, and more directly shows this process than the hodograph does, where the forecaster must rely on bulk shear estimates. The top (cool shades) of the Storm Slinky, where well-removed from the remainder of the profile, indicates good separation and is conducive to long-lived storm modes.

Supercell environments will produce a Storm Slinky profile resembling a kidney bean (Blumberg, 2017), while single-cell convection produces a tightly clustered Storm Slinky where the rings stack on top of each other and exhibit little curvature. The forecaster, where possible, should consider the sensitivity of the Storm Slinky shape to different combinations of parcel lift method and storm motion method.

Since the storm slinky samples the environment above the LFC, it should not be used for characterizing the inflow layer, where storm-relative helicity is criticial.

## A2.6. Equiv. potential temp. profile

Immediately to the right of the storm slinky window is an X/Y graph depicting the equivalent potential temperature ($\theta_e$) profile. The vertical range of this graph spans 600 mb (about 14,000 ft MSL) to the surface. On this graph the equivalent potential temperature is plotted on the abscissa, and is labeled in Kelvin.

This is a very useful graph for instantly identifying the layers which are most supportive of buoyancy. As the lowest layers are not "compressed" like the logarithmic profile of the Skew T, this allows detials in the lowest layers to stand out.

A profile that extends to the right at higher altitudes may indicate nocturnal cooling, as the higher layers decouple from the boundary layer. However a profile that extends significantly to the right at higher altitudes usually indicates an elevated moisture situation, and is typically associated with stations located behind fronts.

## A2.7. Storm-relative wind profile

Another graph appears immediately to the right of the storm slinky and the equivalent potential temperature graph: the storm-relative wind profile. This plots the magnitude of the storm-relative wind at each level, using the Bunkers right-movement vector for storm motion. The vertical range of the scale is fixed from the surface to 15 km AGL (about 49,000 ft), and the abscissa is marked every 10 knots.

Several layer means may appear on this graph. A red vertical bar shows storm-relative wind velocity in the 0-2 km layer, with a blue vertical bar for the 4-6 km layer, and a purple bar for the 9-11 km layer.

Storm-relative anvil-layer shear has been found to be a good discriminator of supercell mode (Rasmussen and Straka 1998). The 40 kt threshold is permanently marked as a dashed purple line near the top of the graph to differentiate low-precipitation (LP) from classic (CL) modes. If the 9-11 km mean SR wind falls to the left of this line, LP modes are more likely if the wind falls to the right. Another purple line far to the left is marked at 70 kt, differentiating classic from low-precipitation (LP) modes. It should be noted that the original literature suggested 60 kt as a threshold for the CL-LP transition, with 20-40 kt for the LP-CL transition.

The importance of the 9-11 km storm-relative wind lies in the separation between the updraft and the precipitation cascade. High storm-relative wind in the upper portions of the thunderstorm indicate good separation and long-lived cells, leading to the steady-state LP supercell modes. Low storm-relative winds tend to disrupt the updraft, but in high instability or high-shear regimes, rainy low-precipitation supercells and rain-wrapped tornadoes may occur.

A dashed vertical white line at the 15 kt point appears. This is used to assess 4-6 km storm-relative wind and helps discriminate between weak and strong mid-level winds. If the 4-6 km SR wind exceeds 15 kt, falling to the right of the white line, then supercells are more likely to be tornadic. However it is likely that the differences here between tornadic and nontornadic supercells are not very well forecast by this parameter (Thompson et al. 2003, and Markowski et al. 2003).

## A2.8. Possible Hazard Type

In SHARPpy, below the hodograph and to the right is a box with the abbreviation for Possible Hazard Type (PHT). This shows the results of an algorithm that considers a wide range of parameters, including Significant Tornado Parameter, effective storm relative helicity, CAPE, precipitable water, and many other variables. It has been developed from these predictors and climatological datasets, and are summarized in Figure X-XX.

As the chart shows, this is not strictly a thunderstorm predictor. It also identifies a variety of other weather types, including blizzard, flash flood, and excessive heat. If tornado potential is forecast, it discriminates between a marginal, regular, and PDS tornado risk. If no tornado risk is identified but thunderstorms are likely, the risk may be classified as severe or marginal severe. If neither tornadoes nor severe thunderstorms are identified, the algorithm will look for flash flood, blizzard, and excessive heat risks, and if these aren't found, then the PHT result will be "none".

It must be stressed that PHT is an amalgamation of various forecast parameters, and has not been fully tested. It is never to be used alone, and it does not forecast air mass

**Table X-XX. Possible Hazard Type (PHT) decision process.** This table outlines the algorithm for the SHARPpy PHT algorithm, proceeding from the top and evaluating each row. All parameters must be true unless there is an "or" condition.

### PDS Tornado
- STPE$\geq$3; STPF$\geq$3; SRH$_{0-1}$$\geq$200 m$^2$s$^{-2}$; ESRH$\geq$200 m$^2$s$^{-2}$; SRW$_{4-6km}$$\geq$15 kt; BWD$_{0-8km}$$\geq$45 kt; LCL$_{SB}$<1000 m; LCL$_{ML}$ < 1200 m; LR$_{0-1}$$\geq$5 C° km$^{-1}$; CINH$_{ML}$>50 J kg$^{-1}$; EIB=0 m

### Tornado (any of 3 conditions)
- STPE$\geq$3 or STPF$\geq$4; CINH$_{ML}$>-125 J kg$^{-1}$; EIB=0 m, **or**
- STPE$\geq$1 or STPF$\geq$1; SRW$_{4-6}$$\geq$15 kt or BWD$_{0-8}$$\geq$40 kt; CINH$_{ML}$>-50 J kg$^{-1}$; EIB=0 m, **OR**
- STPE$\geq$1 or STPF$\geq$1; [(LowRH+MidRH)/2]$\geq$60%; LR$_{0-1}$$\geq$5 C° km$^{-1}$; CINH$_{ML}$>-50 J kg$^{-1}$; EIB=0 m

### Marginal Tornado (any of 2 conditions)
- STPE$\geq$1 or STPF$\geq$1; CINH$_{ML}$>-150 J kg$^{-1}$; EIB=0 m, **OR**
- (STPE$\geq$1 and ESRH$\geq$150 m$^2$s$^{-2}$) or (STPF$\geq$0.5 and SRH$_{0-1}$$\geq$150 m$^2$s$^{-2}$); CINH$_{ML}$>-50 J kg$^{-1}$; EIB=0 m

### Severe (any of 3 conditions)
- STPF$\geq$1 or SCP$\geq$4 or STPE$\geq$1; CINH$_{MU}$$\geq$-50 J kg$^{-1}$, **OR**
- (SCP$\geq$2 and SHIP$\geq$1) or DCAPE$\geq$750 J kg$^{-1}$; CINH$_{MU}$$\geq$-50 J kg$^{-1}$, **OR**
- SigSvr$\geq$30,000 m$^3$s$^{-3}$; MMP$\geq$0.6; CINH$_{MU}$$\geq$-50 J kg$^{-1}$

### Marginal Severe
- WNDG$\geq$0.5 or SHIP$\geq$0.5 or SCP$\geq$0.5; CINH$_{MU}$$\geq$-75 J kg$^{-1}$

### Flash Flood
- PW>Climatological u+2o; Corfidi Upshear Speed < 25 kt

### Blizzard
- Surface wind > 35 kt; Surface temperature $\leq$ 0°C; Bourgouin algorithm indicates snow

### Excessive Heat
- Surface heat index > 105°F

### KEY
BWD - Bulk wind difference, kt; shear between the top and bottom of the subscripted layer
CINH - Convective inhibition, J kg$^{-1}$; lifting method in subscript
Corfidi Upshear Speed - Backbuilding estimate for MCS storms (Corfidi et al 1996) (Corfidi 2003)
EIB - Effective inflow base, m AGL
ESRH - Effective storm-relative helicity, m$^2$s$^{-2}$
LCL - Lifted condensation level, m; lifting method in subscript
LR - Lapse rate, C° km$^{-1}$
STPE - Significant Tornado Parameter, using effective-layer quantities (Thompson et al 2012)
STPF - Signficant Tornado Parameter, using fixed-layer quantities (Thompson et al 2003)
SCP - Supercell Composite Parameter (q.v.)
SHIP - Significant Hail Parameter (q.v.)
SRH - Storm-relative helicity, m$^2$s$^{-2}$
SRW - Storm-relative wind, kt, through km AGL layer subscripted

changes unless the diagram source is specifically from model output. PHT is designed strictly as a check value.

The optimal forecast process is for the forecaster to compose a prediction of the expected weather using soundings, surface data, satellite imagery, model forecasts, and other tools. When that process is complete, then the PHT value is considered. If there is a difference, this is a signal for the forecaster to look at the data again and try to resolve the discrepancy. This can be done by exploring all the different parameters on the sounding, especially the STP output. If the forecaster's prediction matches the PHT result, then this indicates there is strong meteorological support for the prediction.

## A2.9. Thermodynamic output

The first SHARPpy index box contains a very useful summary of thermodynamic information relating to the environment, as well as a listing of parcel characteristics and other tabulations. Virtual temperature is always used where possible. Information on how to using the data in a forecast process are given in the preceding chapters or the suggested references.

### A2.9.1. Instability: summary columns (Section 1)

Instability parameters are given in six columns, with four types of available parcel lifts described in the next section. The six types of instability data are as follows:

* **Convective availability of potential energy (CAPE)** ($J\ kg^{-1}$). This is an integrated type of stability index and provides the total available energy to a parcel.
* **Convective inhibition (CINH, CIN)** ($j\ kg^{-1}$). Indicates the amount of capping. Although CIN is a negative energy expression, SHARPpy plots them as negative numbers. Values of ≥0 indicate a lack of capping, and <–150 is associated with very strong capping.
* **Lifted condensation level height (LCL)**, m. Tornadic storms are more likely with LCL heights of ≤1000 m.
* **Lifted index (LI)**, C°. A lifted index of ≤ –5° is associated with tornadoes and severe weather.

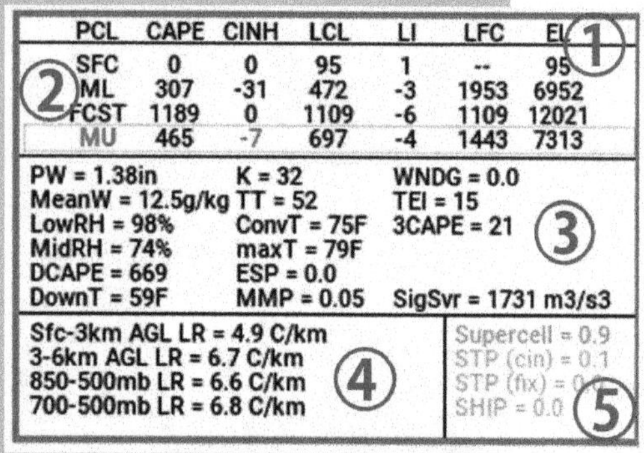

| PCL | CAPE | CINH | LCL | LI | LFC | EL |
|---|---|---|---|---|---|---|
| SFC | 0 | 0 | 95 | 1 | -- | 95 |
| ML | 307 | -31 | 472 | -3 | 1953 | 6952 |
| FCST | 1189 | 0 | 1109 | -6 | 1109 | 12021 |
| MU | 465 | -7 | 697 | -4 | 1443 | 7313 |

PW = 1.38in    K = 32    WNDG = 0.0
MeanW = 12.5g/kg    TT = 52    TEI = 15
LowRH = 98%    ConvT = 75F    3CAPE = 21
MidRH = 74%    maxT = 79F
DCAPE = 669    ESP = 0.0
DownT = 59F    MMP = 0.05    SigSvr = 1731 m3/s3

Sfc-3km AGL LR = 4.9 C/km
3-6km AGL LR = 6.7 C/km
850-500mb LR = 6.6 C/km
700-500mb LR = 6.8 C/km

Supercell = 0.9
STP (cin) = 0.1
STP (fix) = 0
SHIP = 0.0

**Figure A2-3. SHARPpy thermodynamics summary sections**, as described in the text.
1 - Instability: summary columns
2 - Instability: parcel method rows
3 - Additional parameter summary
4 - Lapse rate summary
5 - SCP/STP/SHIP summary

* **Level of free convection (LFC)**, m. An LFC of less than 2500 m is associated with tornadoes.

* **Equilibrium level (EL)**, m. The EL is not a predictor of severe weather, but together with LFC and CAPE it can determine NCAPE. The EL is also useful for identifying low-topped supercell environments.

### A2.9.2. Instability: parcel method rows (Section 2)

There are four rows in the top instability table. These indicate instability parameters based on four different types of parcel lift. Thus, the row ML and the column CAPE gives MLCAPE, or mixed layer CAPE. The four possible rows are as follows:

* **SFC, surface based.** This uses the surface temperature and dewpoint. It is important to note that this represents conditions at observation time, and does not consider the effects of mixing, so its use is generally not recommended.

* **ML, 100 mb mixed layer.** This mixes the lowest 100 mb of the sounding by obtaining the mean potential temperature ($\theta$) of the temperature profile, and the mean mixing ratio ($q$) of the dewpoint profile.

* **FCST, forecast.** This is intended for a morning sounding, computing the effects of surface heating using the data on the sounding. For temperature, $T_{max}$, the 850 mb temperature is mixed dry adiabatically to the surface, and a 2 C° superadiabatic surface layer is added.

* **MU, most unstable.** An equivalent potential temperature ($\theta_e$, theta-e) profile is generated from the surface to 400 mb, and the level with the highest theta-e is selected. The temperature and dewpoint at this level provides the parcel conditions. This method is strongly recommended for elevated convection, especially poleward of frontal systems, and is useful as a "worst case" perspective.

The MU method is used to determine LCL, LFC, and EL on the diagram. This information is plotted at the appropriate heights in green (LCL), yellow (LFC), and purple (EL).

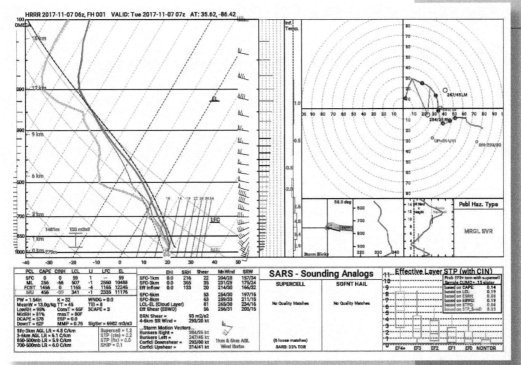

Figure A2-4. SHARPpy full panel view. This was a sample in Tennessee ahead of a line of weak cool season storms; note the 17°C dewpoints near the surface. The local time is 1 a.m. CST and there is a lack of surface heating along with a radiational inversion, thus the CAPE shows zero. However the MUCAPE value shows 456 J kg⁻¹. The precipitable water is also high for December, with 1.54". *(11/7/2017 07Z BNA)*

## A2.9.3. Additional parameter summary (Section 3)

The middle section contains various instability and other miscellaneous parameters as follows:

* **PW** – precipitable water, inches. An expression of the total amount of water vapor in the atmospheric column.

* **MeanW** – Mean mixing ratio, g kg$^{-1}$. This uses the lowest 100 mb of the atmosphere (i.e. the ML parcel).

* **LowRH** – Low-level RH, percent. This is based on the layer from 0–150 mb above the ground.

* **MidRH** – Mid-level RH, percent. This is based on the layer 150–350 mb above the ground.

* **DCAPE** – , J kg$^{-1}$. Downdraft CAPE, a measure of downburst and high wind potential.

* **DownT** – Downrush temperature, °F. This provides the temperature at the ground using the DCAPE parcel.

* **K** – K Index. The legacy instability index as defined by George 1960.

* **TT** – Total Totals Index. The legacy instability index as defined by Miller 1967.

* **ConvT** – Convective temperature, °F. Determines the lowest surface temperature that results in less than 5 J kg$^{-1}$ of CIN using a 100 mb ML parcel.

* **MaxT** – Maximum temperature, °F. Predicts the maximum temperature by using a parcel brought dry adiabatically from 100 mb AGL and adding 2 °C to model near-surface heat (a superadiabatic contact layer proxy).
* **ESP** – Enhanced stretching potential. This is a dynamic index that is proportional to steep lapse rates and CAPE in the lowest 3 km. It is associated with tornado potential.
* **MMP** – MCS maintenance probability, %. An SPC composite parameter which considers deep shear, lapse rate, and mean wind. The value indicates the percent chance an MCS will be strong and well maintained. See Coniglio 2007.
* **WNDG** – Wind damage parameter, nondimensional. Identifies damaging wind potential using MLCAPE, lapse rates, mid-level mean wind, and convective inhibition. Values greater than 1 indicate high wind.
* **TEI** – theta-e index. This is the difference between surface and minimum theta-e value in the lowest 400 mb, and can be used for downdraft forecasting.
* **3CAPE** – 0 to 3 km CAPE. This is simply the MLCAPE in the lowest 3 km.
* **SigSvr** – Significant severe parameter, $m^3 s^{-3}$. This is the product of MLCAPE and 0-6 km shear, providing a value similar to EHI. Over 10,000 indicates severe weather, with significant tornadoes at over 30,000. See Craven and Brooks 2004.

### A2.9.4. Lapse rate summary (Section 4)

The bottom left box contains lapse rate (LR) information. This is not obtained from parcel calculations but rather from the existing environmental profile. Lapse rates of less than approximately 6 C° km$^{-1}$ are considered to be stable and not conducive to vertical motion. Lapse rates between roughly 6 C° km$^{-1}$ and 10 C° km$^{-1}$ are considered to be unstable, and are conducive to buoyancy of saturated parcels. Lapse rates close to 10 C° km$^{-1}$ are dry adiabatic, close to the limit which can exist in the atmosphere, and indicate strong instability and the potential for strong vertical motion.

* **Sfc-3km AGL lapse rate**. This layer is close to the ground, in the lowest 10,000 feet, and high lapse rates here indicate strong surface heating and possibly the advection of a polar air mass which is undergoing strong modification.

* **3-6 km AGL lapse rate.** This is the mid-level lapse rate, between 10,000 and 20,000 feet AGL. High lapse rates here are indicative of strong dynamic lift or differential advection.
* **850-500 mb lapse rate.** This samples a relatively deep layer, from about 5,000 to 18,000 ft MSL. High lapse rates indicate strong dynamic lift and possibly strong thermal advection associated with the movement of a new mass into the region.
* **700-500 mb lapse rate.** This is similar to the 3-6 km AGL lapse rate, sampling the mid-levels of the troposphere, and is used in the same manner.

### A2.9.5. SCP/STP/SHIP summary (Section 5)

1.1.1. Miscellaneous indexes are provided in the lower right of the first SHARPpy output box, and are printed in brown.

* **Supercell – Supercell Composite Parameter (SCP).** Provides SCP (Thompson et al 2007).
* **STP (cin) – Significant Tornado Parameter (CIN).** Provides STP parameter using effective-layer quantities (Thompson et al 2012).
* **STP (fix) – Significant Tornado Parameter (fix).** Provides STP parameter using fixed layer quantities (Thompson et al 2003)
* **SHIP – Significant Hail Prediction.** Hail index, developed internally at SPC. Values greater than 1 indicate that significant hail (2-inch or larger) is likely.

## A2.10. Shear output

The second large SHARPpy parameter output box is dedicated to shear measurements.

### A2.10.1. Shear table (columns)

The following are the five key quantities given by SHARPpy: EHI, SRH, shear, mean wind, and storm-relative wind. They are described as follows:

* **Energy-Helicity Index (EHI).** This is a product of CAPE and helicity, and forecasts the potential for rotating storms. The effective inflow layer and the SFC-1 km layer are the most useful layers.

**Hail size chart**

Below are the official conversion figures used by SPC to assess hail size.

| Dia., Inches | Object |
|---|---|
| 0.50 | Marble, moth ball |
| 0.75 | Penny |
| 0.88 | Nickle |
| 1.00 | Quarter |
| 1.25 | Half dollar |
| 1.50 | Ping pong ball |
| 1.75 | Golf ball |
| 2.00 | Hen egg |
| 2.50 | Tennis ball |
| 2.75 | Baseball |
| 3.00 | Tea cup |
| 4.00 | Softball |
| 4.50 | Grapefruit |

* **Storm relative helicity (SRH)**, $m^2\ s^{-2}$. This indicates the presence of streamwise vorticity being ingested into updraft, and is proportional to the potential for rotating storms with a well-developed mesocyclone.
* **Shear**, kt. This indicates shear between the bottom and top of the layer. It is not an expression of total shear.
* **Mean wind**, deg, kt. Provides mean pressure-weighted wind within the layer, relative to the ground.
* **SRW(storm-relative wind)**, deg, kt. Provides mean pressure-weighted wind within the layer, relative to the storm (right mover, RM). The effective inflow layer and the SFC-1 km layer are especially important; the magnitude equals storm-relative inflow and is proportional to the storm severity. The direction indicates where it is fed from; i.e. 180°/20 is an inflow feed from the south at 20 kt.

## A2.10.2. Shear table (rows)

The rows of the shear table are used to specify different layers in conjunction with the shear and wind data listed above. Here is what each layer is used for:

* **Surface-1 km AGL layer**. This is typically the best available layer representing inflow for a surface-based storm. Effective inflow should be used if there is any doubt about the layer, or if elevated convection is occurring.
* **Surface-3 km AGL layer**. This considers the lowest 10,000 ft and it has only limited use.
* **Effective inflow layer**. The effective inflow layer is the best estimate of layer in all conditions, however the forecaster must check the sounding and make sure this layer is representative.
* **Surface-6 km AGL layer**. This is the deep layer containing most of the storm. The main piece of useful information is shear (yielding bulk shear) and mean wind (which can give some information on storm motion).
* **Surface-8 km AGL layer**. This is mostly useful for storm motion, obtained from the mean wind.
* **Lifted condensation level (LCL) to equilibrium level (EL)** layer, or the "cloud layer". This is called the "cloud layer" since the base of the cumulonimbus cloud is assumed to be the LCL, with the EL forming the anvil top.
* **Effective shear layer (EBWD)**. The effective bulk wind difference is the layer from the effective inflow base to 50%

of the EL height. Stronger shear values (more than 35 kt) are associated with tornadoes, with significant tornadoes having a mean value of 48 kt.

### A2.10.3. Additional shear and wind data

The parameters are given in the five columns, showing the values for the layer identified in each row. All directions are relative to true north:

* **BRN Shear**, $m^{-2}\ s^{-2}$. Gives the Bulk Richardson Number.
* **4-6 km storm-relative wind**, deg, kt. This evaluates the mid-level winds. Some older references suggest that low 4-6 km shear values may be associated with high-precipitation supercells.
* **Bunkers right motion vector**, deg, kt. This provides the best estimate of storm motion using the Bunkers (internal dynamics, ID) method. The value given is ground-relative.
* **Bunkers left motion vector**, deg, kt. Similar to above but for left movers, including left splits. Left motion magnitude that is almost the same as the right motion is often associated with splitting tendencies.
* **Corfidi downshear**, deg, kt. Indicates total storm motion for a forward propagating storm. Low-level inflow is added to the mean wind.
* **Corfidi upshear**, deg, kt. Indicates total storm motion for a backbuilding storm. Low-level inflow is subtracted from the mean wind.

### A2.10.4. Wind barb diagram

In the lower right side of this box, two wind barbs graphically depict the shear by plotting the 1 km AGL wind (red) and 6 km AGL wind (blue). These wind barbs are ground-relative. This gives an instantaneous glance of winds at the inflow layer (1 km) and storm top (6 km). The greater the difference, the higher the bulk shear and the greater the potential for long-lived cells.

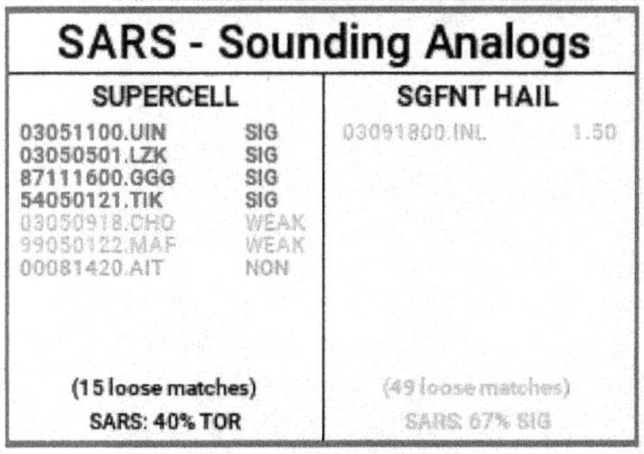

## A2.11. Sounding analogs (SARS)

The Sounding Analog Retrieval System, or SARS (Jewell 2010) produces probabilistic forecasts of *significant hail* and

**Figure A2-5. An example of SARS sounding analog results**, showing matches with previous severe weather events. Note the color coding

### SARS sounding analogs

These are the quantities matched by SHARPpy to obtain matches for supercells and hail.

**Supercell**

The soundings are matched using MLCAPE, ML mixing ratio, 0-1 km shear, 0-1 km SRH, 500 mb temperature, and the 700-500 mb lapse rate. A second pass requires tighter adherence to the values above, and adds 0-3 km shear, 0-3 km SRH, and 0-9 km shear.

**Significant hail**

Matches are performed using MUCAPE, MU mixing ratio, 700-500 mb lapse rate, 500 mb temperature, 0-3 km shear, 0-6 km shear, and 0-9 km shear. The second pass uses closer matches of all the above values, typically doubling the sensitivity, and adds evaluation of 0-3 km helicity. This requires closer adherence to the analog sounding value.

---

*supercells* by comparing current *computed parameters* to computed parameters from a database of 1148 previous soundings between 1989 and 2006 where these severe weather events occurred, most events being on the Great Plains. The algorithm lists the date, time, sounding, and hazard type for which a match was found.

The construction of the SARS database used checks to make sure appropriate soundings were being used. Soundings in the SARS database must have been launched with 100 miles of a hail report, within the same air mass comprising the inflow into the storm, and must have occurred within a 2.5-hour window centered on 2330 UTC in order to try to obtain surface-based storms. The surface data is checked and some soundings are modified.

The SARS algorithm does not match the soundings graphically, but uses extracted quantities. These differ between the supercell and hail algorithms.

### A.2.11.1. SUPERCELL: Supercell analogs tab

The supercell tab on the left attempts to match known tornadic and nontornadic supercell storms. Matches are rated SIG for significant tornadoes, WEAK for minor tornadoes, and NON for nontornadic supercells. Matches in the WEAK and SIG categories are counted as positive for tornadoes, and count toward a probability of 100%, while the NON cases count toward 0%. A table in the side margins provides the parameters which are evaluated for supercell analogs.

### A.2.11.2. SGFNT HAIL: Significant hail analogs tab

The significant hail analog panel forecasts the probability of hail of 2 inches (over golf ball size) and greater. Along with this it determines the most likely maximum hail size, in inches, ranging from 0.75 to over 4 inches. As with the supercell section, matches of 2 inches or greater count toward a probability of 100%, and matches to smaller hail to 0%.

A table in the side margins provides the parameters which are evaluated for hail analogs.

## A2.12. Significant Tornado Parameter (STP)

In the lower right side of the page is significant tornado parameter (STP). A graph here shows tornado intensity on the abscissa (x-axis), from EF4+ down to nontornadic; the ordinate (y-axis) shows the STP numerical value.

There is a green box and whiskers ranking for each tornado intensity category. *These are not observed values* but are reference values drawn from climatological studies. They are fixed with respect to the diagram, thus we see that EF3 tornadoes are associated with a mean STP of about 2.5.

The diagram is read by observing the position of the thick horizontal line. In a fair weather environment it will be at the bottom of the diagram, indicating an STP of 0. In a severe weather situation, this horizontal line will rise up into the diagram. The forecaster observes the position of the line with respect to each tornado intensity category, and this shows the probable distribution of forecast tornado intensity; for instance at STP 7 there should be EF4 and EF5 tornadoes with a few EF3 tornadoes. This is a general guide and tornadoes other than the rankings shown may develop.

A box at the top right lists the parameters that make up the STP calculation. When the STP is elevated, this list is very useful in identifying which meteorological factor is contributing to the STP value. Colors range from brown (weak), to white (marginal), yellow (moderate), and red (strong).

* **Based on CAPE**. This shows the frequency of significant tornadoes based on CAPE. Values ≥2500 J kg$^{-1}$ will be colored yellow. No red is used.
* **Based on LCL**. Indicates the frequency of significant tornadoes based on lifted condensation level; values below 1000 m are colored yellow.
* **Based on ESRH**. Indicates the frequency of significant tornadoes based on effective layer storm relative helicity. Values ≥400 are yellow and ≥500 red.
* **Based on EBWD**. Indicates the frequency of significant tornadoes based on effective layer bulk wind difference. Values ≥50 kt are yellow and ≥70 kt red.
* **Based on STPC**. Indicates the frequency of significant tornadoes based on effective layer significant tornado parameter (STP). Values ≥2 are yellow and ≥4 red, with ≥8 magenta.
* **Based on STP_fixed**. Shows the frequency of significant tornadoes based on fixed layer significant tornado parameter (STP). Values ≥2 are yellow and ≥5 red, with ≥7 magenta.

**Figure A2-6. A listing of the internal data** available to SHARPpy in a SARS (Sounding Analog Retrieval System) sounding. Only the Parcel Information below is used.

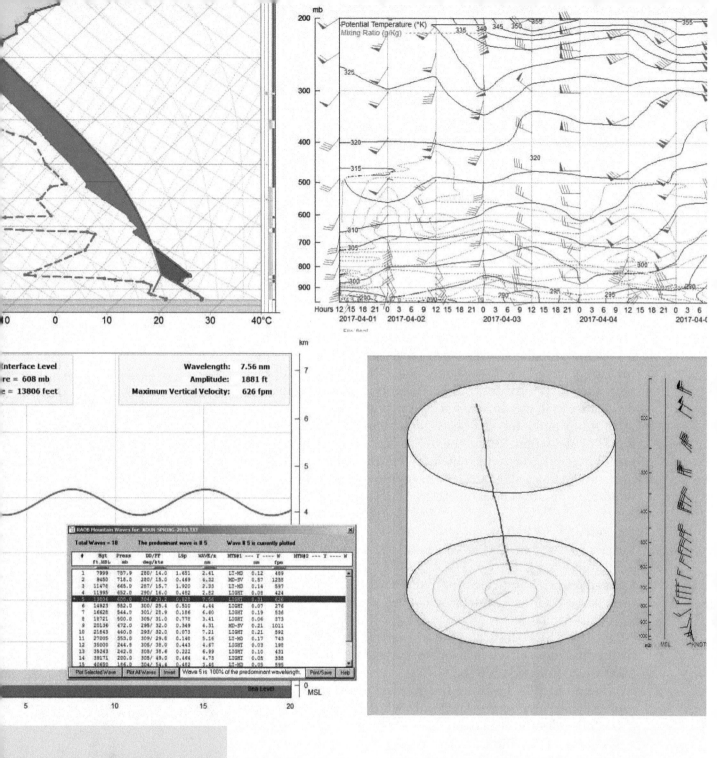

# A-3 RAOB

RAOB is a commercial software package produced by Eosonde Research Services (formerly Environmental Research Services), or ERS, of The Villages, Florida. It specializes in detailed soundings, hodographs, cross sections, mountain wave analysis, soaring parameter analysis, and other functionality relating to radiosonde observations.

RAOB was initially developed in 1987 by John Shewchuk, a retired Air Force weather officer. The first version was programmed on a Tandy 1000 computer and GW Basic, using information found in AWS/TR-79/006, *Equations and Algorithms for Meteorological Applications*.

The next chapter came in 1992 when Shewchuk developed a partnership with Richard Cale, a retired private sector meteorologist. His aircraft icing and mountain wave algorithms up to then had been developed for the HP-41CX programmable calculator. Shewchuk and Cale worked together to implement these routines in RAOB, and this led to a greatly improved software package.

The program was sold to the public beginning in 1995 and shifted from MS-DOS to Windows in 2002. Over the two decades has has worked its way into National Weather Service offices, foreign weather services, universities, military units, and even the NTSB. In the chapter that follows we will provide a brief overview of the software and focus on functions that are useful to severe weather forecasters.

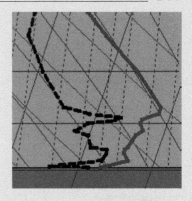

**RAOB author**
RAOB is a Windows executable program that was developed in 1987 and has been continuously upgraded and revised to this day. It provides sounding, hodograph, cross section, 10-100 mb sounding extensions, and many other capabilities.

**Developer**
John Shewchuk

**Program location**
www.weathergraphics.com/raob *
www.raob.com

* author has a financial interest

## A3.1. Data import

Perhaps one of the most important functions besides displaying data is sourcing the data. RAOB has the capability to import existing upper-air data through the File > Open Sounding menu option, or can download data from public radiosonde data sources on the Internet at File > Download Data.

Once the file has been imported, the user is then asked to select a station. There are two different ways to choose the station: with a station table and with a map. Users can toggle to the map by pressing the "map" button, or can return to the station listing by choosing "listing" above the map.

The station table selection mode offers a "filter", which allows the listing to be constrained by WMO region or the first two digits of the WMO station number (WMO block).

**RAOB capabilities**

RAOB provides the following capabilities. Most of these are provided as modules that are purchased separately.
- Analytic module
- Soundingram/scanner
- Hodograph/interactive
- Cross section
- Turbulence & wave
- High altitude (10-100 mb)
- Special decoders
- Binary decoders
- Radiometer
- Sodar/Lidar file decoder
- Doppler file decoder
- Aerosol decoder
- Real time processor
- Standard & binary encoders
- Advanced export
- Merge & compositing module

Severe weather forecasters will get the most benefit from the following modules in the list below, in order from highest to lowest priority. The "Forecaster" package will provide all of these except for the merge module.
- Hodograph
- Analytic
- Cross section
- Merge & compositing module
- Special Decoders Module (which provides useful functionality with FSL, AMDAR, and other formats)

For the conterminous US, applicable WMO blocks are 72 and 74; Canada is 71; Alaska is 70; Mexico is 76; and Hawaii is 91.

As mentioned in previous chapters, operational radiosonde data is provided in upper air collective files and posted by various universities and government offices. Most of these sources post data in the WMO FM 35 TEMP format, however it is often questionable whether the newer upper-air data format, BUFR, is included in these files. Frequent users of near realtime upper air data should periodically audit their files to determine if BUFR data is being included, and if not, work with providers of upper air data to ensure this is available. Failing to monitor sources of upper-air data means soundings and hodographs from some regions, such as Germany, will be unavailable.

## A3.2. RAOB major modules

### A3.2.1. Sounding

The display of soundings is relatively straightforward. RAOB offers not only skew-T log-p diagrams, but also the closely-related tephigram, which is popular in Europe, and the older emagram. The skew-T and emagram also offer the ability for the upper limit of the chart to be adjusted, including extending it upward above 100 mb. This book will focus on the skew-T diagram.

Forecasters should make extensive use of the options provided in Options > Diagram Options. This allows extensive customization of the diagram appearance, line widths, and colors.

The Analyses Data tab allows information to be displayed on the diagram giving LFC, CCL, SWEAT index, lifted index, total totals, and more. This may be plotted at different corners of the chart. The forecaster can also opt to limit the data to only CAPE, SRH, EHI, VGP, MVV, and BRN, all of which are the basic buoyancy and shear parameters, with various ratios and derivations of those values.

It can be very useful to restrict the top and bottom limit of the diagram. This allows a sort of zoom functionality so that the lower layers can be analyzed in more detail, particularly in capping situations where parcels are very sensitive to the correct starting temperature and dewpoint. This is changed by going to Pres & Temp tab under Upper Limit and

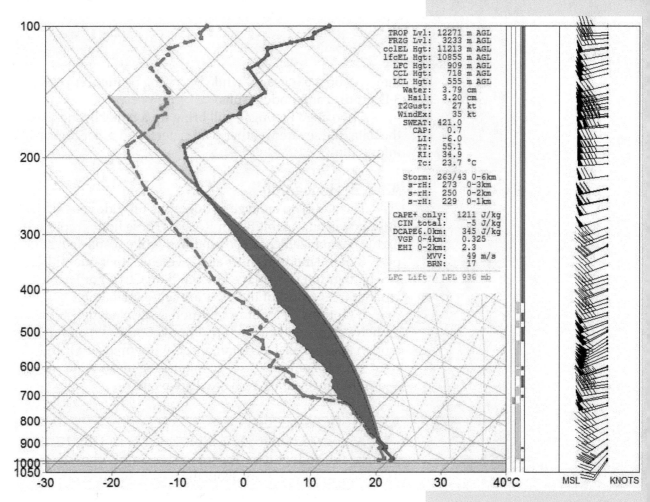

Figure A3-1. **Nashville 1998 tornado sounding** at 1700 UTC, about three hours before an F3 tornado that hit the downtown area at 2030 UTC. This is plotted in RAOB with the basic sounding and analytic module.

Lower Limit. Users can also constrain the horizontal scaling by going to Temperature / Scaling: Manual and setting an arbitrary maximum and minimum temperature corresponding to the desired left and right side temperature at 1000 mb. Users should at least familiarize theirselves with this capability as changing the appearance of the diagram is a very important skill and it exploits the capabilities offered by RAOB.

There is also a Program Options panel under "Options > Program Options" for settings that affect RAOB system-wide. This includes the option to use virtual temperature, the units system, screen colors, storm motion, algorithm depth, CAPE options, layers.

A play and forward/reverse button appears at the top left, allowing the forecaster to tab through the available soundings in the dataset, if there are any. This is useful for datasets with multiple dates/times or multiple stations. The Sounding Scanner (shown in Figure 3-20) plots all of the soundings in the dataset simultaneously as a Stuve diagram with a large assortment of instability, severity, and shear parameters on a bar graph, while Sounding Sequencer produces 3-D profiles and hodographs.

Choosing "Listings" will bring up a panel listing all of the data, the severe weather parameters, and other forecast information.

### A3.2.2. Hodograph

The ability to analyze hodographs provides one of the most powerful capabilities of RAOB. These hodographs are highly customizable, can be modified while displayed, and plot all of the modern shear and severe weather quantities. The hodograph is displayed by pressing Ctrl+H or Display > Hodographs. Most of the plots are self-explanatory.

If Options > Plot Storm Motion is selected, the storm motion vectors will be plotted. For the Bunkers method, this motion appears as left (gray), mean wind (magenta), and right (gray). For the 30R75 technique, the mean wind (olive) and right mover (dark red) are displayed).

### A3.2.3. Cross section

A cross section module is provided in RAOB. To perform the hodographs, the forecaster simply chooses File > Create A Cross Section. Here the user is prompted for the type of diagram. Either a time section may be plotted (with increasing time from left to right on the x-axis) or a distance section (increasing distance along the path on the x-axis). Choose either.

The program then shows a list box, where the stations can be added: one station and date-time per row. Click "Add A Sounding" to add rows. The user will be prompted to choose a file, and then the stations.

The data will them be displayed. Various parameter can be plotted or removed, and gradient fills like that in Figure X-XX can be used.

## A3.5. Other features

Numerous other features beyond the scope of this book are available in RAOB. These include a fire weather plot, soundingram, layer analyzer, frontal analysis module, mountain wave analysis, soaring forecasts, a Google Earth interface, and special modules for interfacing with commercial upper air observation systems.

**Figure A3-2: RAOB hodograph window**, immediately after it is selected. This shows a north-westerly flow configuration, with the right motion shear vector indicating a motion of 120/52 and the left motion vector 090/28. This is a configuration favoring left mover dominance. Also note the SR anvil level shading (blue/gray circles at top left); this configuration is different from UCAR in that it is assessed against the sounding origin (0,0), not the hodograph profile. *(4/12/1994 00Z AMA)*

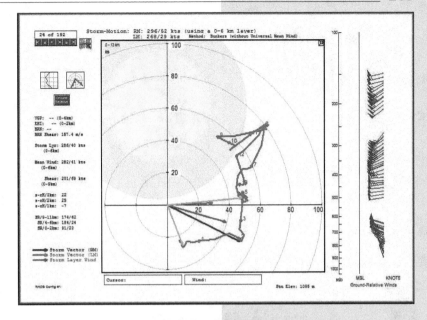

**Figure A3-3: One of the early versions of the RAOB software program**, RAOB 4.0, which ran under MS-DOS. *(Courtesy John Shewchuk)*

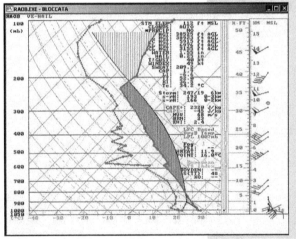

**Figure A3-4. Cross section analysis leading up to the "polar vortex storm"** during the first nine days of January 2014 in at Joliet, Illinois. The color gradient is wind speed and the solid lines are potential temperature (theta). This shows the oscillations between a warm and cold atmosphere. Where the isentropes form "hills" this indicates cold conditions. It also shows a close relationship between upper-level wind direction and the temperature of the lower half of the troposphere, with only slight departures out of phase.

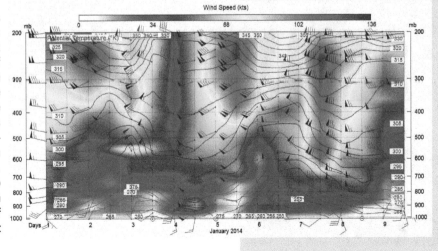

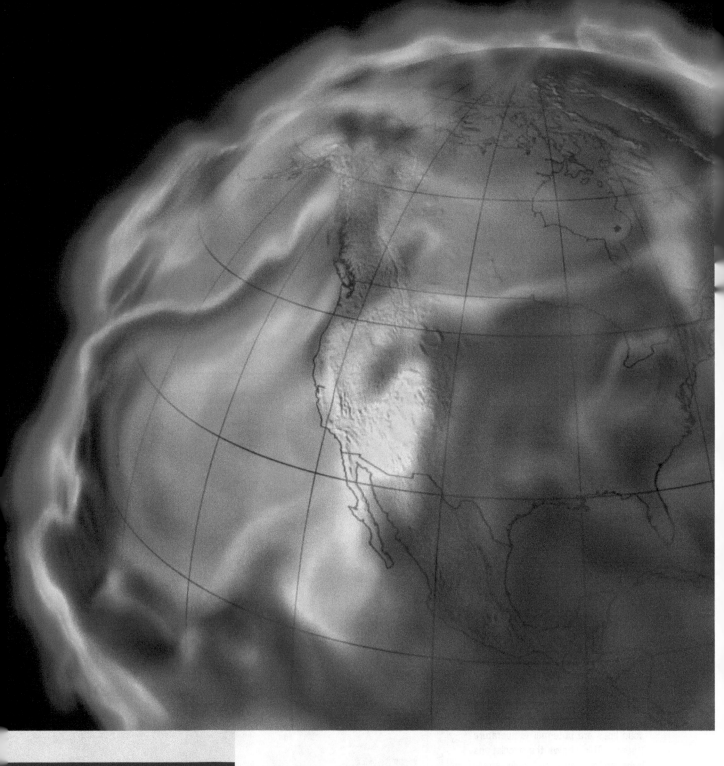

# A-4  OTHER RESOURCES

OTHER RESOURCES 211

Provided here is a listing of thermodynamic data from other sources on the Internet.

## A4.1. University of Wyoming

The University of Wyoming offers atmospheric science degree programs <www.uwyo.edu/atsc> and operates a King Air aircraft for research and a supercomputer center.

The department operates a weather homepage with an upper air website < http://weather.uwyo.edu/upperair/>. It provides extensive upper air analysis capabilities, including analysis of levels above 100 mb up to 10 m. A description of the abbreviations and their units are shown below.

University of Wyoming's weather page provides comprehensive upper air analysis displays, including PDF files.

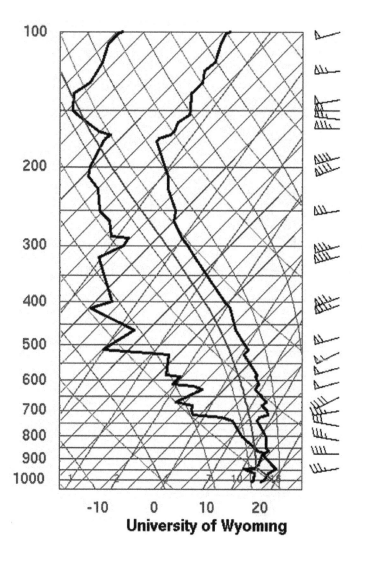

**University of Wyoming**

| | | |
|---|---|---|
| SLAT | 30.50 | SLAT - Latitude (decimal degrees) |
| SLON | -81.70 | SLON - Longitude (decimal degrees) |
| SELV | 9.00 | SELV - Elevation (m) |
| SHOW | 2.14 | SHOW - Showalter Index (°C) |
| LIFT | 3.41 | LIFT - Lifted index (T) (°C) |
| LFTV | 3.02 | LFTV - Lifted index ($T_v$) (°C) |
| SWET | 279.7 | SWET - SWEAT Index |
| KINX | 20.10 | KINX - K Index |
| CTOT | 20.50 | CTOT - Cross Totals Index |
| VTOT | 22.70 | VTOT - Vertical Totals Index |
| TOTL | 43.20 | TOTL - Total Totals Index |
| CAPE | 0.00 | CAPE - CAPE (T) (J $kg^{-1}$) (ML 100 mb) |
| CAPV | 0.00 | CAPV - CAPE ($T_v$) (J $kg^{-1}$) (ML 100 mb) |
| CINS | 0.00 | CINS - Convective inhibition (T) (J $kg^{-1}$)* |
| CINV | 0.00 | CINV - Convective inhibition ($T_v$) (J $kg^{-1}$)* |
| EQLV | -9999 | EQLV - Equilibrium level (T) (mb)* |
| EQTV | -9999 | EQTV - Equilibrium level ($T_v$) (mb)* |
| LFCT | -9999 | LFCT - Level of free convection (T) (mb) |
| LFCV | -9999 | LFCV - Level of free convection ($T_v$) (mb) |
| BRCH | 0.00 | BRCH - Bulk Richardson Number (T) |
| BRCV | 0.00 | BRCV - Bulk Richardson Number ($T_v$) |
| LCLT | 288.5 | LCLT - Lifted condensation level (K) |
| LCLP | 941.6 | LCLP - Lifted condensation level (mb) |
| MLTH | 293.6 | MLTH - Mixed layer potential temp (K) |
| MLMR | 11.86 | MLMR - Mixed layer mixing ratio (g $kg^{-1}$) |
| THCK | 5889. | THCK - Thickness 1000-500 mb (m) |
| PWAT | 33.94 | PWAT - Precipitable water (mm) |

NOTES:
T - Parcel computed with air temp.
$T_v$ - Parcel computed with virtual temp.
* - ML 100 mb for all quantities above

The skew-T diagram shown here has been cropped to fit on the page.

*(University of Wyoming)*

## A4.2. UCAR weather

The University Corporation for Atmospheric Research (UCAR) is an association made up of over 70 universities with doctoral programs in the atmospheric sciences. Its goal is to enhance the computing abilities of individual universities and to carry out scientific missions that are larger than a single university can sustain.

The National Center for Atmospheric Research (NCAR) is federally funded and operated by UCAR. It offers a weather page <weather.rap.ucar.edu> that was one of the first comprehensive real-time weather homepages on the Internet, and still operates to this day. The soundings are found under the Upper Air page.

The Research Applications Laboratory, part of National Center for Atmospheric Research, operates the UCAR weather page, offering soundings (below). Here the various instability, moisture, and severe parameters are explained below. Some quantities are not specified on the website and can be difficult to determine without further investigation, such as CAPE parcel method, but they are provided here.

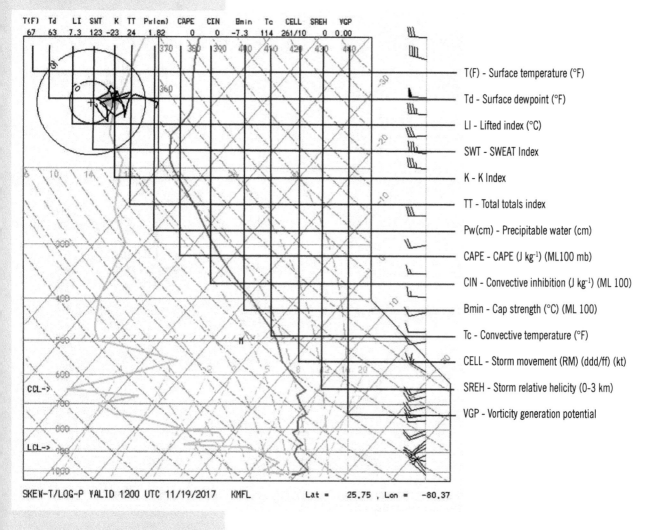

T(F) - Surface temperature (°F)
Td - Surface dewpoint (°F)
LI - Lifted index (°C)
SWT - SWEAT Index
K - K Index
TT - Total totals index
Pw(cm) - Precipitable water (cm)
CAPE - CAPE (J kg$^{-1}$) (ML100 mb)
CIN - Convective inhibition (J kg$^{-1}$) (ML 100)
Bmin - Cap strength (°C) (ML 100)
Tc - Convective temperature (°F)
CELL - Storm movement (RM) (ddd/ff) (kt)
SREH - Storm relative helicity (0-3 km)
VGP - Vorticity generation potential

# A4.3. Storm Prediction Center

The Storm Prediction Center (SPC), the national convective weather prediction center for NOAA, operates an extensive experimental weather page < http://www.spc.noaa.gov/exper/>. This provides many different tools for severe weather forecasters, including sounding climatology.

Of special interest is the Observed Sounding Analysis page at < www.spc.noaa.gov/exper/soundings/>. This allows soundings to be constructed using the SHARPpy interface described in Chapter 7. As a result, it provides data using thoroughly tested algorithms, extensive hodograph analysis, and useful tools like SARS.

It is important to be aware that this works only on observed radiosonde data, and certain parameters will not be valid at the time of maximum heating. It may be more useful in some cases to use Pivotal Weather to view model soundings.

An example of an SPC observed sounding showing northwest flow in Louisiana. A loaded gun configuration is shown, but the atmosphere was too stable for severe weather. Note that significant helicity is present in the 0 – 2 km layer.

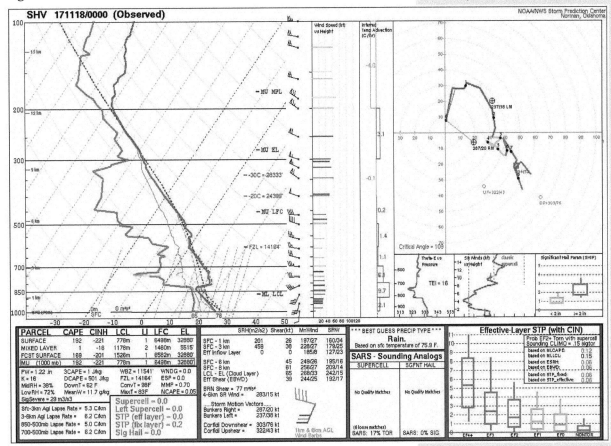

## A4.4. PivotalWeather.com

PivotalWeather is a noteworthy weather forecasting website. It is a project developed in 2015 by alumni of the University of Oklahoma meteorology program. It offers many displays of model data, and the wide range forecast soundings are one of the strongest features.

The forecast soundings are found by going to any of the model data and clicking on the map at the desired point. It may be necessary to enable the soundings by going to the top right and choosing "Soundings" and making sure they are "on". If they are grayed out and read "off", this means that forecast soundings are not available.

The forecast soundings are available for the GFS, NAM, RAP, and HRRR models. They may not be available in some zoomed windows. They are not available with the CFS, nor in most of the foreign models like the ECMWF, GDPS, and RDPS. They are also not available in the HRW displays, nor in the research models like the NSSL WRF.

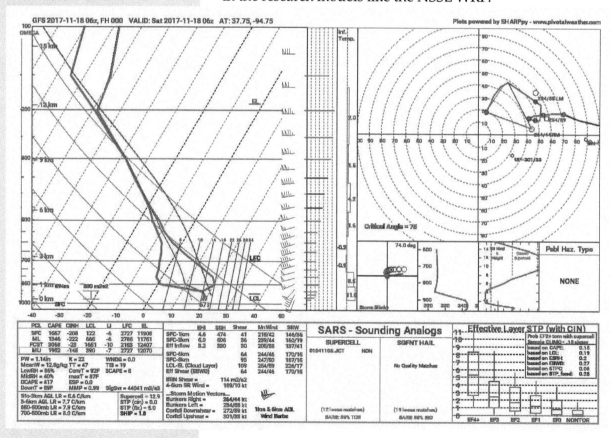

PivotalWeather.com GFS forecast for southwest Kansas in an active weather situation showing a capping inversion with rich low-level moisture. The ability to query mesoscale models and sample their structure vertically is a powerful tool for weather forecasters. PivotalWeather is one of the only sites that allow model data to be queried like this; the other is the NOAA ESRL sounding site at: <rucsoundings.noaa.gov>.

## A4.5. ESRL RUC Soundings site

The NOAA ESRL RUC Soundings site <rucsoundings.noaa.gov> was developed by NOAA physicist and web developer Bill Moninger in 2006. The page was designed to provide forecasters with a way of viewing model data in the vertical using soundings.

As of 2017 the website provides forecast soundings from the RAP (Rapid Refresh) model, which is the same model used by Storm Prediction Center for its experimental mesoanylsis page. It also offers soundings from the FIM, GFS, and NAM model, as well as some internal NCEP models. Profiler data was available until 2014 and has been memoved due to the shutdown of the National Profiler Network. There is also the ability to view radiosonde data

A powerful feature of the site is the ability to show the highlight the cursor position on the skew-T and hodograph simultaneously. This allows detailed analysis of inversions.

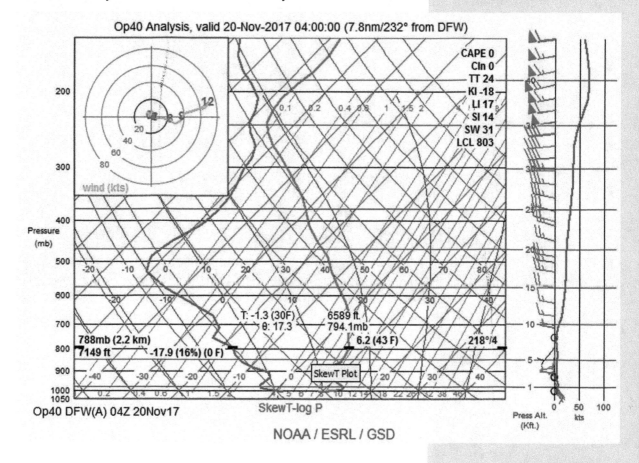

Plot showing forecast data for DFW. There is an interactive sunding and hodograph, with the black line on the sounding and the red mark on the hodograph indicating the cursor location. A wind velocity graph is also shown on the right side.

# A-5 Radiosonde Data

## APPENDIX A-5A
# RADIOSONDE MODELS IN COMMON USE

Frequency given is in MHz. Abbreviations: (D) - Ducted. For a good list of biases, see Schroeder (2003).

| Country | Model | Year | Thermometer | Humidity | Barometer | Freq. | Status/notes |
|---|---|---|---|---|---|---|---|
| AU | Astor/Phillips | 1964 | Rod thermistor | Lithium chloride | Aneroid capsule | 403 | |
| CH | Meteolab. Snow White | 1996 | N/A | Chilled mirror | N/A | 403 | Used as high-precision test instrument |
| CH | Meteolabor SRS-C34 | 2006? | Thermocouple | Hygroclip | Hypsometer | 403 | Used in Switzerland, 2014 |
| CH | Meteolabor SRS-C50 | 2016 | Thermocouple | Capacitive | Hypsometer | 403 | Probably in current use in Switzerland |
| CN | GZZ-2 / GZZ-59 | 1964 | Spiral (D 2x) | Goldbeaters skin | Aneroid capsule | 400 | |
| CN | Shanghai GTS-1 | 1998 | Thermistor, rod | Hygristor, carbon film | Silicon piezores. | 1675 | Used at 75 stations in China, 2014 |
| CN | China Huayun GTS 1-1 | 2000s | Thermistor, bead | Thin film capacitor | Silicon piezores. | 403 | Used at 9 stations in China, 2014 |
| CN | Nanjing GTS1-2 | 2000s | Thermistor (cyl) | Thin film capacitor | Aneroid capsule | 1675 | Used in 10 stations in China, 2014 |
| CN | Beijing CF-06-A | 2000s | Thermistor, bead | Thin film capacitor | Aneroid capsule | 403 | |
| DE | Graw M60 | 1960 | Bimetal cylinder (D) | Artificial hair | Aneroid capsule | 27/43 | |
| DE | Graw RSG | 1978 | Bead thermistor | Psych./carbon hyg. | Aneroid capsule | 403 | |
| DE | Graw Mini. TDFS-87 | 1987 | Bead thermistor | Carbon element | Aneroid capsule | 403 | |
| DE | Graw DFM-97 | 1997 | Capacitive | Thin film capacitor | Aneroid capsule | 403 | Used in India, Mexico 2014 |
| DE | Graw DFM-06 | 2006 | Capacitive | Thin film capacitor | Aneroid capsule | 403 | Used in Iran, 2014 |
| DE | Graw DFM-09 | 2009 | Thermistor | Thin film capacitor | Aneroid capsule | 403 | |
| FI | Vaisala RS11 | 1938 | Bimetal strip | Normal hair / rolled hair | Aneroid capsule | 25 | |
| FI | Vaisala RS12 | 1959 | Bimetal strip (D) | Rolled hair | Aneroid capsule | 25 | |
| FI | Vaisala RS13/15 | 1965 | Bimetal strip | Rolled hair | Aneroid capsule | 25 | |
| FI | Vaisala RS18 | 1976 | Bimetal ring (D) | Rolled hair | Aneroid capsule | 25 | Moderately dry |
| FI | Vaisala RS21 | 1978 | Bimetal ring (D) | Rolled hair | Aneroid cap 2x | Both | Dry bias, more than RS18/RS80 |
| FI | Vaisala RS41-SG | 2013 | Platinum thermistor | Capacitive thin film | GPS | 403 | |
| FI | Vaisala RS80 | 1981 | Capacitive bead | Capacitive thin film | Capacitive an. | Both | Used at one station in Ethiopia, 2014 |
| FI | Vaisala RS90 | 1997 | Capacitive | Capacitive, dual poly. | Silicon piezores. | 403 | Used in New Zealand, Thailand, 2014 |
| FI | Vaisala RS92-SGP | 2003 | Capacitance wire | Thin film capacitor | Silicon piezores. | 403 | Extensive use: Europe, Japan, Australia |
| FR | Modem M2K2DC | 2006 | Bead thermistor | Capacitive polymer | GPS | 403 | Used in South Korea, Malaysia 2014 |
| JP | Meisei RS-01G | 2001 | Rod thermistor | Capacitive polymer | Aneroid capsule | 403 | Used in Indonesia, Japan 2014 |
| JP | Meisei RS-06G | 2006 | Thermistor | Capacitance polymer | Aneroid capsule | 403 | Used in Japan, 2016 |
| JP | Meisei RSII-56 | 1956 | Bimetal cylinder (D) | Hair | Phosphor-bronze | 403 | |
| JP | Meisei RSII-76 | 1976 | Rod thermistor | Carbon hygristor | Ni-span aneroid | 403 | |
| JP | Meisei RSII-80 | 1980 | Rod thermistor | Carbon hygristor | Ni-span aneroid | 403 | |
| JP | Meisei RSII-85 | 1985 | Rod thermistor | Carbon strip | Ni-span aneroid | 403 | |
| KR | Jinyang RSG-20A | 2000s | Thermistor | Thin film capacitor | Capacitive | 403 | |
| RS | Moltchanov RS-049 | 1957 | Bimetal plate/cyl (D) | Hair | Aneroid boxes (2x) | 403 | |
| RS | A-22 | 1957 | Spiral bimetal (D) | Goldbeaters skin | Bronze aneroid | 403 | |
| RS | AVK/MRZ | 1983 | Rod thermistor | Goldbeaters skin | None | 1782 | Used extensively in Russia, 2014 |
| RS | RKZ | 1967 | Rod thermistor | Goldbeaters skin | Baroswitch | 1782 | |
| UK | Kew Mk II | 1946 | Bimetal cylinder (D) | Goldbeaters skin | Aneroid | Unk. | Dry bias |
| UK | Kew Mk III | 1977 | Wire thermistor | Goldbeaters skin | Aneroid capsule | 403 | |
| US | Bendix radiosonde | 1954 | Rod thermistor (D) | Carbon hygristor | Aneroid | 403 | |
| US | Lockheed LMS6 | 2000s | Thermistor (chip) | Thin film capacitor | Capacitive | 403 | Used in the United States after 2014 |
| US | Space Data 909-10-01 | 1988 | Bead thermistor | Carbon hygristor | Aneroid capsule | 1680 | May be wetter than Viz |
| US | VIZ AMT-4 until 1957 | 1954 | Rod thermistor | Lithium chloride | Baroswitch | 1680 | |
| US | VIZ AMT-4 | 1957 | Rod thermistor | Carbon hygristor | Baroswitch | 1680 | |
| US | VIZ 1395 | 1975 | Rod thermistor | Carbon hygristor | Baroswitch | 403 | |
| US | VIZ A (1492-510) | 1986 | Rod thermistor | Carbon hygristor | Baroswitch | 1680 | Moderately wet |
| US | VIZ B (1492-520) | 1988 | Rod thermistor | Carbon hygristor (D) | Baroswitch | 1680 | Moderately wet |
| US | VIZ B2 | 1997 | Rod thermistor | Carbon hygristor | Baroswitch | 1680 | Slightly wet |
| US | VIZ Mk I | 1980 | Rod thermistor | Carbon hygristor | Baroswitch | 403 | |
| US | VIZ Mk II | 1987 | Rod thermistor | Carbon hygristor | Capacitive | Both | Was used in US late 00s/early 10s |
| ZA | iMet-2 | ? | Bead thermistor | Thin film capacitor | Silicon piezores. | 403 | |

## APPENDIX A-5B
# UNITED STATES RADIOSONDE STATUS (Conterminous U.S.)

| WMO | ICAO | STATION NAME | LEGACY DATE | LEGACY EQUIP | RRS DATE | RRS EQUIP | UPGRADE DATE | UPGRADE EQUIP | UPGRADE#2 DATE | UPGRADE#2 EQUIP |
|---|---|---|---|---|---|---|---|---|---|---|
| 72230 | KBMX | AL BIRMINGHAM | 1998-06-01 | RS80 | 2007-03-12 | MIIA | 2013-10-31 | LMS6 | | |
| 72340 | KLZK | AR LITTLE ROCK | 1998-06-01 | RS80 | 2007-02-26 | MIIA | 2013-09-26 | LMS6 | | |
| 72376 | KFGZ | AZ FLAGSTAFF | 1998-06-01 | RS80 | 2006-06-19 | MIIA | 2012-07-17 | RS92 | | |
| 72274 | KTWC | AZ TUCSON | 1995-11-?? | RS80 | 2007-06-04 | MIIA | 2013-04-01 | RS92 | | |
| 72493 | KOAK | CA OAKLAND | 1997-06-01 | SPNB2 | 2009-02-09 | MIIA | 2013-10-10 | LMS6 | | |
| 72293 | KNKX | CA SAN DIEGO | 1997-06-01 | SPNB2 | 2010-04 | MIIA | 2013-05-01 | RS92 | | |
| 72393 | KVBG | CA VANDENBERG AFB | (Department of Defense site) SDC/RT(1998-2002) VIZII/G(2007-2013-) | | | | | | | |
| 72469 | KDEN | CO DENVER | 1995-11-?? | RS80 | 2008-10-27 | MIIA | 2013-12-19 | LMS6 | | |
| 72476 | KGJT | CO GRAND JUNCTION | 1998-06-01 | RS80 | 2006-09-18 | MIIA | 2013-10-31 | LMS6 | | |
| 72206 | KJAX | FL JACKSONVILLE | 1998-06-01 | RS80 | 2007-01-15 | MIIA | 2013-05-01 | RS92 | | |
| 72201 | KEYW | FL KEY WEST | 1997-06-01 | SPNB2 | 2010-03 | MIIA | 2012-06-26 | RS92 | | |
| 72202 | KMFL | FL MIAMI | 1998-06-01 | RS80 | 2007-05-07 | MIIA | 2013-04-01 | RS92 | | |
| 72214 | KTLH | FL TALLAHASEE | 1998-06-01 | RS80 | 2006-03-13 | MIIA | 2013-10-31 | LMS6 | | |
| 72210 | KTBW | FL TAMPA BAY | 1998-06-01 | RS80 | 2007-03-26 | MIIA | 2013-10-31 | LMS6 | | |
| 72215 | KFFC | GA PEACHTREE (ATLANTA) | 1998-06-01 | RS80 | 2008-04-21 | MIIA | 2013-10-22 | LMS6 | | |
| 74455 | KDVN | IA DAVENPORT | 1998-06-01 | RS80 | 2007-11-05 | MIIA | 2013-12-19 | LMS6 | | |
| 72681 | KBOI | ID BOISE | 1995-11-?? | RS80 | 2006-01-27 | MIIA | 2013-12-19 | LMS6 | | |
| 74560 | KILX | IL LINCOLN | 1998-06-01 | RS80 | 2007-07-30 | MIIA | 2013-08-08 | LMS6 | | |
| 72451 | KDDC | KS DODGE CITY | 1997-06-01 | SPNB2 | 2010-06 | MIIA | 2013 | LMS6 | | |
| 72456 | KTOP | KS TOPEKA | 1998-06-01 | RS80 | 2009-04-06 | MIIA | 2013-10-09 | LMS6 | | |
| 72240 | KLCH | LA LAKE CHARLES | 1998-06-01 | RS80 | 2006-03-27 | MIIA | 2013-04-01 | RS92 | | |
| 72233 | KLIX | LA SLIDELL | 1998-06-01 | RS80 | 2008-03-03 | MIIA | 2013-04-01 | RS92 | | |
| 72248 | KSHV | LA SHREVEPORT | 1998-06-01 | RS80 | 2006-04-10 | MIIA | 2013-09-10 | LMS6 | | |
| 74494 | KCHH | MA CHATHAM | 1998-06-01 | RS80 | 2009-04-27 | MIIA | 2014-02-05 | LMS6 | | |
| 72712 | KCAR | ME CARIBOU | 1997-06-01 | SPNB2 | 2012-06 | MIIA | 2013 | RS92 | | |
| 72501 | KGYX | ME GRAY | 1998-06-01 | RS80 | 2006-10 | M11A | 2013-04-01 | RS92 | | |
| 72632 | KDTX | MI WHITE LAKE | 1998-06-01 | RS80 | 2009-05-11 | MIIA | 2013-05-01 | RS92 | | |
| 72634 | KAPX | MI GAYLORD | 1998-06-01 | RS80 | 2007-08-27 | MIIA | 2013-03-01 | RS92 | | |
| 72649 | KMPX | MN CHANHASSEN | 1998-06-01 | RS80 | 2005-09-07 | MIIA | 2013-09-05 | LMS6 | | |
| 72747 | KINL | MN INTERNATIONAL FALLS | 1995-11-?? | RS80 | 2009-05-25 | MIIA | 2013-05-01 | RS92 | | |
| 72440 | KSGF | MO SPRINGFIELD | 1998-06-01 | RS80 | 2007-08-13 | MIIA | 2012-07-10 | RS92 | | |
| 72235 | KJAN | MS JACKSON | 1997-06-01 | RS80 | 2008-03-17 | MIIA | 2013-10-31 | LMS6 | | |
| 72768 | KGGW | MT GLASGOW | 1997-06-01 | SPNB2 | 2007-07-17 | MIIA | 2013-10-31 | LMS6 | | |
| 72776 | KTFX | MT GREAT FALLS | 1997-06-01 | SPNB2 | 2010-05 | MIIA | 2013-04-01 | RS92 | | |
| 72317 | KGSO | NC GREENSBORO | 1998-06-01 | RS80 | 2007-02-12 | MIIA | 2013-11-20 | LMS6 | | |
| 72305 | KMHX | NC MOREHEAD CITY | 1998-06-01 | RS80 | 2007-01-29 | MIIA | 2013-03-01 | RS92 | | |
| 72764 | KBIS | ND BISMARCK | 1997-06-01 | SPNB2 | 2006-08-21 | MIIA | 2013-10-31 | LMS6 | | |
| 72562 | KLBF | NE NORTH PLATTE | 1998-06-01 | RS80 | 2007-09-10 | MIIA | 2013-10-31 | LMS6 | | |
| 72558 | KOAX | NE VALLEY (OMAHA) | 1998-06-01 | RS80 | 2007-09-24 | MIIA | 2013-10-23 | LMS6 | | |
| 72365 | KABQ | NM ALBUQUERQUE | 1997-06-01 | SPNB2 | 2008-05-05 | MIIA | 2013-08-15 | LMS6 | | |
| 72364 | KEPZ | NM SANTA TERESA (ELP) | 1995-12-?? | RS80 | 2008-05-19 | MIIA | 2013-03-01 | RS92 | | |
| 72582 | KLKN | NV ELKO | 1997-06-01 | SPNB2 | 2007-06-18 | MIIA | 2013-10-31 | LMS6 | | |
| 72388 | KVEF | NV LAS VEGAS | ????-??-?? | ???? | ????-??-?? | MIIA | 2013-12-12 | LMS6 | | |
| 72489 | KREV | NV RENO | 1998-06-01 | RS80 | 2006-07-03 | MIIA | 2013-08-08 | LMS6 | | |
| 72518 | KALY | NY ALBANY | 1997-05-20 | SPNB2 | 2009-05-25 | MIIA | 2013-10-01 | LMS6 | | |
| 72528 | KBUF | NY BUFFALO | 1998-06-01 | RS80 | 2006-10-16 | MIIA | 2013-08-19 | LMS6 | | |
| 72501 | KOKX | NY UPTON | 1997-05-20 | MIIA | 2009-06 | MIIA | 2013-04-01 | RS92 | | |
| 72426 | KILN | OH WILMINGTON | 1998-06-01 | RS80 | 2007-04-09 | MIIA | 2013-10-17 | LMS6 | | |
| 72357 | KOUN | OK NORMAN (OKC) | 1998-06-01 | RS80 | 2006-11-13 | MIIA | 2013-10-24 | LMS6 | | |
| 74646 | KLMN | OK LAMONT | (Dept of Energy site) RS90/L(1998-2008) RS92(2009-2017) (Some IMET-1 2014-2016) | | | | | | | |
| 72597 | KMFR | OR MEDFORD | 1997-06-01 | SPNB2 | 2009-03-16 | MIIA | 2013-03-01 | RS92 | | |
| 72694 | KSLE | OR SALEM | 1997-06-01 | SPNB2 | 2007-07-02 | MIIA | 2013-10-31 | LMS6 | | |
| 72520 | KPBZ | PA PITTSBURGH | 1997-06-01 | SPNB2 | 2009-05 | MIIA | 2013-10-07 | LMS6 | | |
| 72208 | KCHS | SC CHARLESTON | 1997-06-01 | SPNB2 | 2008-06-23 | MIIA | 2014-03-20 | LMS6 | | |
| 72662 | KUNR | SD RAPID CITY | 1998-06-01 | RS80 | 2006-07-24 | MIIA | 2013-10-31 | LMS6 | | |
| 72659 | KABR | SD ABERDEEN | 1998-06-01 | RS80 | 2006-08-07 | MIIA | 2013-10-03 | LMS6 | | |
| 72327 | KBNA | TN NASHVILLE | 1997-06-01 | SPNB2 | 2007-05-21 | MIIA | 2013-12-12 | LMS6 | | |
| 72363 | KAMA | TX AMARILLO | 1995-11-?? | RS80 | 2006-05-08 | MIIA | 2013-12-11 | LMS6 | | |
| 72250 | KBRO | TX BROWNSVILLE | 1997-06-01 | SPNB2 | 2009-02 | MIIA | 2013-04-01 | RS92 | | |
| 72251 | KCRP | TX CORPUS CHRISTI | 1997-06-01 | SPNB2 | 2005-10-04 | MIIA | 2013-05-01 | RS92 | | |
| 72261 | KDRT | TX DEL RIO | 1997-06-01 | SPNB2 | 2008-06-09 | MIIA | 2013-12-11 | LMS6 | | |
| 72249 | KFWD | TX FORT WORTH | 1998-06-01 | RS80 | 2007-12-10 | MIIA | 2013 | LMS6 | | |
| 72265 | KMAF | TX MIDLAND | 1995-11-?? | RS80 | 2006-04-24 | MIIA | 2013-10-31 | LMS6 | | |
| 72318 | KRNK | VA BLACKSBURG | 1998-06-01 | RS80 | 2007-02-26 | MIIA | 2013-11-25 | LMS6 | | |
| 72403 | KIAD | VA STERLING (DC) | 1995-11-01 | RS80 | 2005-08-01 | LMS5 | 2016-09 | LMG6* | | |
| 72402 | KWAL | VA WALLOPS ISLAND | 1996-07-24 | MIIA | - | | - | | | |
| 72797 | KUIL | WA QUILLAYUTE | 1997-06-01 | SPNB2 | 2007-10-22 | MIIA | 2013-05-01 | RS92 | | |
| 72786 | KOTX | WA SPOKANE-WA | 1997-06-01 | SPNB2 | 2006-05-22 | MIIA | 2013-09-12 | LMS6 | | |
| 72645 | KGRB | WI GREEN BAY | 1997-06-01 | SPNB2 | 2008-10-13 | MIIA | 2013-10-31 | LMS6 | | |
| 72672 | KRIW | WY RIVERTON | 1995-11-?? | RS80 | 2006-09-05 | MIIA | 2013-10-31 | LMS6 | | |

## APPENDIX A-5C
## UNITED STATES RADIOSONDE STATUS (Alaska and Hawaii)

| WMO | ICAO | STATION NAME | LEGACY DATE | LEGACY EQUIP | RRS DATE | RRS EQUIP | UPGRADE DATE | UPGRADE EQUIP | UPGRADE#2 DATE | UPGRADE#2 EQUIP |
|---|---|---|---|---|---|---|---|---|---|---|
| 70273 | PANC | AK ANCHORAGE | 1998-06-01 | RS80 | 2008-07-21 | MIIA | | | 2016-09 | LMG6 |
| 70398 | PANT | AK ANNETTE | 1998-10-30 | SPNB2 | 2009-09 | MIIA | 2013-09-13 | LMS6 | | |
| 70026 | PABR | AK BARROW | 1998-12-04 | SPNB2 | 2012-08 | MIIA | | | | |
| 70219 | PABE | AK BETHEL | 1995-11-?? | RS80 | 2010-08 | MIIA | 2013-09-28 | LMS6 | | |
| 70316 | PACO | AK COLD BAY | 1995-12-?? | RS80 | 2010-04 | SPNB2 | | | | |
| 70261 | PAFA | AK FAIRBANKS | 1995-11-?? | RS80 | 2009-08 | MIIA | 2013-09-09 | LMS6 | | |
| 70326 | PAKN | AK KING SALMON | 1995-11-?? | RS80 | 2009-07 | MIIA | 2013-09-10 | LMS6 | | |
| 70350 | PADQ | AK KODIAK | 1995-11-?? | RS80 | 2009-01 | SPNB2 | 2013-09-25 | LMG6* | | |
| 70133 | PAOT | AK KOTZEBUE | 1995-11-?? | RS80 | 2010-04 | SPNB2 | | | | |
| 70231 | PAMC | AK MCGRATH | 1995-11-?? | RS80 | 2009-11 | SPNB2 | | | | |
| 70200 | PAOM | AK NOME | 1995-11-?? | RS80 | 2009-07 | MIIA | 2013-09-09 | LMS6 | | |
| 70308 | PASN | AK ST PAUL ISLAND | 1998-12-10 | SPNB2 | 2010-08 | MIIA | 2013-09-27 | LMS6 | | |
| 70361 | PAYA | AK YAKUTAT | 1995-11-?? | RS80 | 2009-08 | MIIA | 2013-09-24 | LMS6 | | |
| 91285 | PHTO | HI HILO | 1998-11-18 | SPNB2 | 2010-07 | MIIA | 2014-12-02 | LMS6 | | |
| 91165 | PHLI | HI LIHUE | 1995-11-01 | RS80 | 2010-07 | MIIA | 2014-12-18 | LMS6 | | |

## APPENDIX A-5D
## CANADA RADIOSONDE STATUS

| WMO | ICAO | STATION NAME | LEGACY DATE | LEGACY EQUIP | RRS DATE | RRS EQUIP |
|---|---|---|---|---|---|---|
| 71119 | CWSE | AB EDMONTON/STONY PLN | 1995-12-?? | RS80 | 2005-12-15 | RS92 |
| 71945 | CYYE | BC FORT NELSON | 1992-12 | RS80 | 2005-10-29 | RS92 |
| 71203 | CYLW | BC KELOWNA | 1992-12 | RS80 | 2005 | RS92 |
| 71109 | CYZT | BC PORT HARDY | 1992-12 | RS80 | 2005 | RS92 |
| 71908 | CZXS | BC PRINCE GEORGE 71896 | 1992-12 | RS80 | 2005 | RS92 |
| 71913 | CYYQ | MB CHURCHILL | 1992-12 | RS80 | 2005 | RS92 |
| 71867 | CYQD | MB THE PAS | 1996-03 | VIZW9000 | 2005 | RS92 |
| 71816 | CYYR | NL GOOSE BAY | 1995-12 | RS80 | 2005 | RS92 |
| 71801 | CYYT | NL ST JOHNS | 1999-06 | VIZW9000 | 2005 | RS92 |
| 71815 | CYJT | NL STEPHENVILLE | 1992-12 | VALCOM | 2005 | RS92 |
| 71600 | CYSA | NS SABLE ISLAND | 1995-12 | RS80 | 2007-03-29 | RS92 |
| 71603 | CYQI | NS YARMOUTH | 1992-12 | VIZ B | 2005 | RS92 |
| 71934 | CYSM | NT FORT SMITH | 1995-12 | RS80 | 2005 | RS92 |
| 71957 | CYEV | NT INUVIK | 1997-11 | RS80 | 2005 | RS92 |
| 71043 | CYVQ | NT NORMAN WELLS | 1992-12 | VALCOM | 2005 | RS92 |
| 71082 | CWLT | NU ALERT | 1995-12 | RS80 | 2005 | RS92 |
| 71926 | CYBK | NU BAKER LAKE | 1994-02 | RS80 | 2005 | RS92 |
| 71925 | CYCB | NU CAMBRIDGE BAY | 1992-12 | VALCOM | 2005 | RS92 |
| 71915 | CYZS | NU CORAL HARBOUR | 1992-12 | VIZ B | 2005 | RS92 |
| 71917 | CWEU | NU EUREKA | 1992-12 | VALCOM | 2005 | RS92 |
| 71081 | CYUX | NU HALL BEACH | 1992-12 | RS80 | 2006-05-29 | RS92 |
| 71909 | CYFB | NU IQALUIT | 1995-12 | RS80 | 2005 | RS92 |
| 71924 | CYRB | NU RESOLUTE BAY | 1995-12 | RS80 | 2005 | RS92 |
| 71836 | CYMO | ON MOOSONEE | 1998-?? | RS80? | 2005 | RS92 |
| 71845 | CYPL | ON PICKLE LAKE | 1992-12 | VIZ B | 2005 | RS92 |
| 71306 | CYPH | QC INUKJUAQ (71907) | 1992-12 | VALCOM | 2005 | RS92 |
| 71906 | CYVP | QC KUUJJUAQ | 1992-12 | VALCOM | 2005 | RS92 |
| 71823 | CYAH | QC LA GRANDE IV | 1992-12 | VALCOM | 2005-11-25 | RS92 |
| 71722 | CWMW | QC MANIWAKI | 1992-12 | RS80 | 2005 | RS92 |
| 71811 | CYZV | QC SEPT ILES | 1992-12 | VALCOM | 2005 | RS92 |
| 71964 | CYXY | YT WHITEHORSE | 1994-09 | RS80 | 2005 | RS92 |

# A-6 GLOSSARY

**absolute humidity** · Water vapor content of the air in g m$^{-3}$ or kg$^{-1}$.
**absolute temperature** · Temperature relative to absolute zero.
**absolute zero** · The lowest possible temperature: 0 K, -273.15 °C.
**adiabatic process** · Heating and cooling without any external source of heat.
**adiabat** · A lapse rate line representing adiabatic heating and cooling of a parcel of air.
**air** · Air in the Earth's atmosphere consists primarily of nitrogen and oxygen, with contributions from argon and a large number of trace gases.
**BRCH** · (GEMPAK) Bulk Richardson number (q.v.) using air temperature
**BRCV** · (GEMPAK) Bulk Richardson number (q.v.) using virtual temperature
**buoyancy** · The upward or downward force exerted on a volume due to its displacement of a surrounding volume with different density.
**CAPE** · Convective available potential energy (buoyancy)
**CAPV** · (GEMPAK) Convective available potential energy (q.v.) using virtual temperature
**CINS** · (GEMPAK) Convective inhibition using air temperature
**CINV** · (GEMPAK) Convective inhibition (q.v.) using virtual temperature
**Clausius-Clapeyron equation** · An equation that allows the vapor pressure to be calculated.
**condensation** · Phase transition from gas to liquid. This process releases latent heat into the parcel, causing heating.
**conditional instability** · A state where instability exists if the air parcel is saturated, but is stable if the parcel is not saturated.
**conduction** · The transfer of heat by direct contact.
**convection** · The vertical transfer of heat through buoyancy.
**convective available potential energy** · A measure of the amount of energy available for convection. It can be related to the parcel's vertical velocity.
**convective inhibition** · A measure of the energy suppressing upward motion due to the existence of a relatively warm layer (an inversion).
**cross totals index** · An early severe weather index developed by Col. Robert C. Miller which relates upper moisture quantities to lapse rate.
**CTOT** · (GEMPAK) Cross totals index (q.v.)
**density** · the ratio between the mass of a substance and its volume
**dewpoint temperature** · The temperature at which cooling would result in saturation and condensation.

**diabatic** · An adjective describing processes in which external energy is transferred, such as air mass changes from cold ground or warm ocean surfaces.
**dry adiabatic** · A temperature change involving a parcel that is not saturated.
**dry air** · In a thermodynamic process, a parcel without water vapor.
**EQLV** · (GEMPAK) equilibrium level (q.v.) using air temperature
**EQTV** · (GEMPAK) equilibrium level (q.v.) using virtual temperature
**energy** · A property that allows work to be done, such as heating or motion.
**equivalent potential temperature** · The temperature that would result from all latent heat being released into a parcel, and the parcel being moved dry adiabatically to 1000 mb to provide a common reference level.
**evaporation** · Phase transition from liquid to gas. This absorbs latent heat from the air, causing cooling.
**gas** · The highest energy phase of ordinary matter, in which the molecules are widely dispersed, with variable shape and volume.
**heat** · the flow of energy from a hot object to a cold one
**heterogeneous nucleation** · The typical method by which condensation develops in the atmosphere on condensation nuclei, which includes sulfates, nitrates, salt, dust, pollen, etc.
**homogenous nucleation** · The spontaneous condensation of moisture without condensation nuclei; typically this requires temperatures below -40 °C.
**humidity** · A general term referring to the presence of water vapor. Definitions include relative humidity, specific humidity, and absolute humidity.
**inversion** · A layer of air in which the lapse rate shows a significant decrease in cooling, or shows warming with height.
**isobar** · A line of equal pressure.
**isotherm** · A line of equal temperature.
**isentrope** · A line of equal potential temperature.
**Kelvin** · Units of temperature equivalent to Celsius, using absolute zero as a base (0 K = -273.15°C).
**KINX** · (GEMPAK) K-Index (q.v.)
**LAPS** · (GEMPAK) lapse rate (q.v.)
**lapse rate** · The change in temperature per change in height (distance); can be given as °C km$^{-1}$ or K m$^{-1}$.
**latent heat** · Thermal energy absorbed or released during a phase change. Changes from a lower to higher energy state absorb energy, while changes from higher to lower release energy.
**LCL** · lifted condensation level (q.v.)

**LCLP** · (GEMPAK) lifted condensation level pressure
**LCLT** · (GEMPAK) lifted condensation level temperature
**level of free convection** · The level at which a rising parcel becomes warmer than the surrounding environment.
**LFCT** · (GEMPAK) level of free convection using air temperature
**LFCV** · (GEMPAK) level of free convection using virtual temperature
**LFTV** · (GEMPAK) lifted index using 100 mb mixed layer and virtual air temperature
**LIFT** · (GEMPAK) lifted index using 100 mb mixed layer and air temperature
**lifted condensation level** · P
**liquid** · A phase of matter between the energy state of solid and gas, in which the volume is fixed but the shape is variable.
**mixing ratio** · P
**MLTH** · (GEMPAK) Mean potential temperature, mixed layer
**MLMR** · (GEMPAK) Mean mixing ratio, mixed layer
**moist adiabatic** · P
**moist air** · In a thermodynamic process, a parcel with water vapor.
**nucleation** · P
**parcel** · A volume of air of arbitrary size. For the purposes of this book a parcel can be assumed to be in the range of meters.
**potential temperature** · The temperature a volume of air would have if it was moved to the 1000 mb level at the dry adiabatic rate.
**precipitable water** · The depth of water that would cover the ground if all the water vapor in the column were condensed. It can be given in inches of water depth, or kg m$^{-2}$.
**pseudo-adiabatic** · P
**PWAT** · (GEMPAK) precipitable water (q.v.)
**radiation** · The transfer of heat through electromagnetic energy, typically infrared imagery but can include visible light, ultraviolet light, X-rays, gamma rays, and radio waves.
**relative humidity** · The ratio of the existing vapor pressure in a volume to the saturation vapor pressure, in which water vapor and condensed water are in equilibrium.
**RICH** (GEMPAK) Richardson number.
**SELV** · (GEMPAK) Station elevation.
**SHOW** · (GEMPAK) Showalter stability index.
**SHRD** · (GEMPAK) Wind shear direction.
**SHRM** · (GEMPAK) Wind shear magnitude.
**skew-T log-p diagram** · A thermodynamic diagram developed in 1947 in which the temperature axis is skewed upward and to the right, and isobars are straight. This is the most common type of thermodynamic diagram in the United States.
**SLAT** · (GEMPAK) Station latitude.
**SLON** · (GEMPAK) Station longitude
**solid** · The phase of matter with the lowest energy state; volume and shape are fixed.
**specific heat**
**specific humidity**
**STAB** · (GEMPAK) potential temperature lapse rate
**stability**
**STAP** · (GEMPAK) stability with respect to pressure (K mb$^{-1}$)
**STID** · (GEMPAK) station ICAO code
**STNM** · (GEMPAK) station WMO number
**supercooling** · the lowering of liquid water temperature below its freezing point (0 °C) without it becoming a solid.
**SWEAT index** · Severe Weather Threat index, developed in 1970 by Col. Robert C. Miller. This was one of the first indices specifically for tornado prediction. The second version, introduced in 1971, included a shear term.
**SWET** · (GEMPAK) SWEAT index (q.v.)
**temperature** · A measure of kinetic energy of the constituent molecules making up a volume or a substance.
**tephigram** · A thermodynamic diagram similar to the skew-T log-p diagram, but recognizable by its slightly curved isobars.
**thermodynamic diagram** · A diagram used for performing thermodynamic calculations. Temperature is on the x-axis, with height on the y-axis.
**theta** · Potential temperature (q.v.)
**theta-e** · Equivalent potential temperature (q.v.)
**theta-w** · Wet-bulb temperature (q.v.)
**thickness** · The geopotential distance between two pressure surfaces. It is proportional to the mean virtual temperature of a layer.
**THCK** · thickness (typically 1000-500 mb if not specified)
**TOTL** · (GEMPAK) Total Totals Index
**virtual temperature** · The temperature which the parcel, if devoid of water vapor, would have to heat to in order to have the same density.
**VTOT** · (GEMPAK) Vertical Totals Index
**water vapor** · Water in the gas phase.
**wet adiabat** · Moist adiabat (q.v.)
**wet-bulb temperature** · the lowest temperature that a parcel can achieve through evaporation of water.
**wet-bulb potential temperature** · wet-bulb temperature (q.v.) reduced wet-adiabatically to 1000 mb. Skew-T log-p moist adiabats are labeled in units of wet-bulb potential temperature.

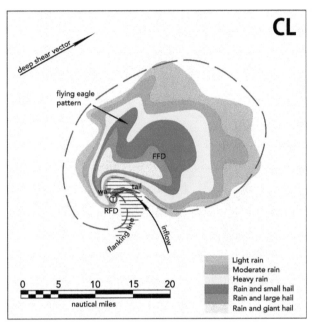
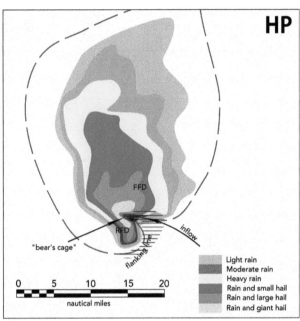
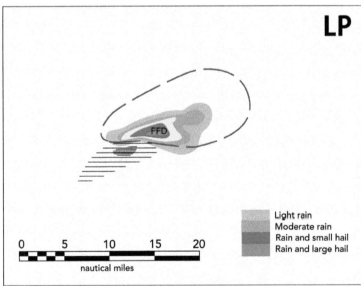

**Figure A1-1. Diagrams of the three supercell archetypes.** The classic (CL) supercell is characterized by a well-defined forward-flank downdraft (FFD), an active rear-flank downdraft (RFD) region, and a strong updraft. The high-precipitation (HP) supercell is identified by the very strong RFD region and the appearance of the updraft on the forward flank of the storm, sometimes much further north than what is indicated here. The low-precipitation (LP) supercell is indicated by a small, persistent FFD region and the absence of an RFD.

A-4

Air Weather Service, 1979: The use of the Skew T, Log P diagram in analysis and forecasting, AWS/TR-79/006, United States Air Force. 153 pp.

American Meteorological Society, cited 2017: Climatology. Glossary of Meteorology. [Available online at http://glossary.ametsoc.org/wiki/]

Atkins, N.T. and R.M. Wakimoto, 1991: Wet Microburst Activity over the Southeastern US: Implications for Forecasting. Wea. Forecasting, 6, 470-482.

Baumgardt, D.A., and K. Cook, 2006: Preliminary evaluation of a parameter to forecast environments conducive to non-mesocyclone tornadogenesis. 23rd Conference on Severe Local Storms, St. Louis, Paper 12.1.

Beebe, R. G., 1958: Tornado proximity soundings. Bull. Amer. Meteor. Soc., 39, 195-201.

Bidner, A., 1970: The Air Force Global Weather Central severe weather threat (SWEAT) index—A preliminary report. Air Weather Service Aerospace Sciences Review, AWS RP 105-2, No. 70-3, 2-5.

Blanchard, D. O., 1998: Assessing the vertical distribution of convective available potential energy. Wea. Forecasting, 13, 870–877.

Bluestein, H.B. and M.L. Weisman, 2000: The interaction of numerically simulated supercells initiated along lines. Mon. Wea. Rev., 128, 3128-3149.

Blumberg, W. G., K. T. Halbert, T. A. Supinie, P. T. Marsh, R. L. Thompson, and J. A. Hart, 2017: "SHARPpy: An Open Source Sounding Analysis Toolkit for the Atmospheric Sciences." Bull. Amer. Meteor. Soc., in press.

Boyden, C. J., 1963: A simple instability index for use as a synoptic parameter. Meteor. Mag., 92, 198–210.

Brown, R. A., 1990: Characteristics of supercell hodographs. Preprints, 16th Conf. on Severe Local Storms, Kananaskis Park, Alta., Canada, Amer. Meteor. Soc., 30-33.

Browning, K. A. and F. H. Ludlam, 1962: Airflow in convective storms. Quart. J. Roy. Meteor. Soc., 88, 117-135.

Bryan, G. H. and J. M. Fritsch (2000) Moist absolute instability: The sixth static stability state. Bull. Amer. Meteor. Soc., 81, 1207—1230.

Bunkers, M. J., B. A. Klimowski, J. W. Zeitler, R. L. Thompson, and M. L. Weisman, 2000: Predicting supercell motion using a new hodograph technique. Wea. Forecasting, 15, 61-79.

_____, et al. 2006: An Observational Examination of Long-Lived Supercells. Part I and II: Characteristics, Evolution and Demise. Wea. Forecasting, 21, 673-688, 689-714.

_____, D. A. Barber, R. L. Thompson, R. Edwards, and J. Garner, 2014: Choosing a universal mean wind for supercell motion prediction. J. Oper. Meteor., 2, 115 –129.

Burgess, D.W., 2003: Supercells. Presentation at the COMAP Course, Boulder, CO.

Carlson, T. N., R. A. Anthes, M. Schwartz, S. G. Benjamin and D. G. Baldwin, 1980: Analysis and prediction of severe storms environment. Bull. Amer. Met. Soc., 61, 1018-1032.

Caruso, J. M., and J. M. Davies, 2005: Tornadoes in non-mesocyclone environments with pre-exisitng vertical vorticity along convergence boundaries. NWA Electronic Journal of Operational Meteorology, June 2005.

Cavender, A., 2014: Analysis of the factors promoting tornadogenesis using NCDC storm events and IGRA data. Thesis, Rosenstiel School of Marine and Atmospheric Science, 71 pp.

Chisholm, A. J., and J. H. Rennick, 1972: The kinematics of multicell and supercell Alberta hailstorms. Alberta hail studies, Research Council of Alberta Hail Studies, Rep. 72-2, 24-31, 53 pp.

Colby Jr., F. P., 1984: Convective inhibition as a predictor of convection during AVE-SESAME

II. Mon. Wea. Rev., 112, 2239–2252.

Coniglio, M. C., H. E. Brooks, S. J. Weiss, and S. F. Corfidi, 2007: Forecasting the maintenance of quasi-linear mesoscale convective systems. Wea. Forecasting, 22, 556-270.

Corfidi, S.F., J.H. Merritt, and J.M. Fritsch, 1996: Predicting the movement of mesoscale convective complexes. Weather and Forecasting, 11, 41-46.

Corfidi, S. F., 2003: Cold pools and MCS propagation: Forecasting the motion of downwind-developing MCSs. Wea. Forecasting, 18, 997–1017.

Craven, J. P., 2000: A Preliminary Look at Deep Layer Shear and Middle Level Lapse Rates Associated with Major Tornado Outbreaks. Preprints, 20th Conference on SLS, Orlando, FL, AMS, 547-550.

_____, and H. E. Brooks, 2004: Baseline climatology of sounding derived parameters associated with deep, moist convection. Nat. Wea. Digest, 28, 13-24. (PDF)

Darkow, G. L., 1968: The total energy environment of severe storms. J. Appl. Met., 7, 199-205.

Davies, J. M., 1993: Hourly helicity, instability, and EHI in forecasting supercell tornadoes. Preprints, 17th Conf. on Severe Local Storms, St. Louis, MO, Amer. Meteor. Soc., 107-111.

Davies, J. M., 2006: Tornadoes in environments with small helicity and/or high LCL heights. Wea. and Forecasting, 21, 579-594.

Davies-Jones, R., 1984: Streamwise vorticity: The origin of updraft rotation in supercell storms. J. Atmos. Sci.,41, 2991–3006.

_____, D. W. Burgess, and M. Foster, 1990: Test of helicity as a tornado forecast parameter. Preprints, 16th Conf. on Severe Local Storms, Kananaskis Park, AB, Canada, Amer. Meteor. Soc., 588-592.

Doswell, C. A. III, and E. N. Rasmussen, 1994: The effect of neglecting the virtual temperature correction on CAPE calculations. Wea. Forecasting, 9, 625-629.

_____, 1991: A review for forecasters on the application of hodographs to forecasting severe thunderstorms. Nat. Wea. Digest, 16 (1), 2-16.

_____, and J. S. Evans, 2003: Proximity sounding analysis for derechos and supercells: An assessment of similarities and differences. Atmos. Research, 67-68, 117-133.

Edwards, R., and R.L. Thompson, 1998: Nationwide comparisons of hail size with WSR-88D vertically integrated liquid water and derived thermodynamic sounding data. Wea. Forecasting, 13, 277-285.

Edwards, R., R. L. Thompson, and C. M. Mead, 2004: Assessment of anticyclonic supercell environments using close proximity soundings from the RUC model. Preprints, 22ndConf. on Severe Local Storms, Hyannis, MA, Amer. Meteor. Soc.

Esterheld, J. M., and D. J. Giuliano, 2008: Discriminating between tornadic and non-tornadic supercells: A new hodograph technique. Electron. J. Severe Storms Meteor., 3 (2). <www.ejssm.org/ojs/index.php/ejssm/article/viewArticle/33>

Evans, J.S., and C.A. Doswell III, 2002: Investigating Derecho and Supercell Proximity Soundings. Preprints, 21st Conf. Severe Local Storms, San Antonio.

Fawbush, E. J. and R. C. Miller, 1952: A mean sounding representative of the tornadic airmass environment. Bull. Amer. Meteor. Soc., 33, 303-307.

_____ and R. C. Miller, 1953: A method for forecasting hailstone size at the earth's surface. Bull. Amer. Meteor. Soc., 34, 235-244.

Fujita, T.T. (1981): Tornadoes and downbursts in the context of generalized planetary scales. J. Atmos. Sci., 38, 1511-1534.

George, J.J., 1960: Weather Forecasting for Aeronautics. Academic Press, New York.

Gilmore, M.S., and L.J. Wicker, 1998: The influence of midtropospheric dryness on supercell morphology and evolution. Wea. Forecasting, 126, 943-958.

Hamilton, William Rowan . "The Hodograph, or a New Method of Expressing in Symbolic Language the Newtonian Law of Attraction", Proceedings of the Royal Irish Academy, Vol. 3 (1847), pp. 344–353.

Hart, J. A., and W. D. Korotky, 1991: The SHARP workstation—v1.50. A skew T/hodograph analysis and research program for the IBM and compatible PC: User's manual. NOAA/NWS Forecast Office, Charleston, WV, 62 pp.

Houston, A. L., R. Thompson and R. Edwards, 2008: The Optimal Bulk Wind Differential Depth and the Utility of Upper-Tropospheric Storm-Relative Flow for Forecasting Supercells. Wea. Forecasting., 23, 825-837.

Hyson, P., R. M. Leigh, and R. L. Southern, 1964: Observational and forecasting aspects of the convection cycle at Darwin. Proc. Symposium on Tropical Meteorology. Rotorua, New Zealand Meteorological Service, 306-315.

Jefferson, G. J., 1963a: A modified instability index. Meteorol. Mag., 92, 92-96.

_____, 1963b: A further development of the instability index. Meteorol. Mag., 92, 313-316.

_____, 1966: Letter to the editor. Meteorol. Mag., 95, 381-382. The third and final revision of the Jefferson Index.

Johns, R. H., and C. A. Doswell III, 1992: Severe local storms forecasting. Wea. Forecasting,7, 588–612.

Johnson, A. W. and K. E. Sugden, 2014: Evaluation of sounding-derived thermodynamic and wind-related parameters associated with large hail events. Electronic J. Severe Storms Meteor., 5 (9), 1-42.

Knight , N. C., and M. English , 1980: Patterns of hailstone embryo type in Alberta hailstorms . J. de Recherches Atmospheriques, 14 , 325-332.

Kutzbach, G., 1979: The Thermal Theory of Cyclones. Amer. Meteor. Soc., 255 pp.

Lilly, D. K., 1986: The structure, energetics and propagation of rotating convective storms. J. Atmos. Sci.,43, 113–140.

Linacre, E., 1992: Climate data and resources. Routledge, London, 367 pp.

Luers, J. K. and R. E. Eskridge, 1998: Use of Radiosonde Temperature Data in Climate Studies. J. Climate, 11, 1002–1019.

Maddox, R. A., 1976: An evaluation of tornado proximity wind and stability data. Mon. Wea. Rev.,104, 133–142.

Markowski, Paul M., E. N. Rasmussen, J. M. Straka, and D. C. Dowell, 1998: Observations of Low-Level Baroclinity Generated by Anvil Shadows. Mon. Wea. Rev.: Vol. 126, No. 11, pp. 2942–2958.

Marwitz, J. D., 1972: The structure and motion of severe hailstorms. Part I: Supercell storms. J. Appl. Meteor., 11, 166–179.

Markowski, P. M., and J. M. Straka, 2000: Some observations of rotating updrafts in low-buoyancy, highly-sheared environments. Mon. Wea. Rev., 128, 449-461.

Miller, D. J., 2006: Observations of low level thermodynamic and wind shear profiles on significant tornado days. Preprints, 23rd Conf. on Severe Local Storms, St. Louis, MO, Amer. Meteor. Soc., 1206-1223.

Miller, R. C., 1967: Notes on analysis and severe storm forecasting procedures of the Military Weather Warning Center. AWS Tech. Rep. 200, USAF, 170 pp. Introduced the Vertical Totals Index.

_____, 1972: Notes on Analysis and Severe-Storm Forecasting Procedures of the Air Force Global Weather Central. AWS Tech. Rpt. 200 (rev). Air Weather Service, Scott AFB, Il 190 pp.

_____, A. Bidner, and R. A. Maddox, 1971: The use of computer products in severe weather forecasting (the SWEAT index). Preprints. Seventh Conf. Severe Local Storms. Kansas City, Amer. Meteor. Soc, 1-6.

\_\_\_\_, 1975: Notes on analysis and severe storm forecasting procedures of the Air Force Global Weather Central. Tech. Report 200.

\_\_\_\_ and R. A. Maddox, 1975: Use of the SWEAT and SPOT indices in operational severe storm forecasting. Preprints. Ninth Conf. Severe Local Storms. Norman, Amer. Meteor. Soc, 1-6.

Moller, A. R., C. A. Doswell, M. P. Foster, and G. R. Woodall, 1994: The operational recognition of supercell thunderstorm environments and storm structures. Wea. Forecasting, 9, 327–347.

Moore, J.T., C.H. Pappas, and F.H. Glass, 1993: Propagation characteristics of mesoscale convective systems. Preprints, 17th Conference on Severe Local Storms, St. Louis (AMS), 538-542.

Moncrieff, M. W. and M. J. Miller, 1976: The dynamics and simulation of tropical cumulonimbus and squall lines. Quart. J. R. Met. Soc., 102, 373-394.

National Weather Service, 1998: Revised specification for barometer, precision digital. National Weather Service Engineering Division Specification G395-SP001, Silver Spring, 18 pp.

\_\_\_\_, 2010: Upper Air Program NWSPD 10-14. Department of Commerce. < http://www.nws.noaa.gov/directives/sym/pd01014001curr.pdf>

Nordahl, L. S., 1978: Evaluation of radiosonde relative humidity (RH). National Weather Service Internal Memorandum, NWS, Silver Spring, MD, 2 pp.

Normand, C. W. B., 1946: Energy in the atmosphere. Quart. J. Roy. Meteor. Soc., 72, 145-167.

Orlanski, I., 1975: A rational subdivision of scales for atmospheric processes. Bull. Am. Meteor. Soc., 56, 527-530.

Patrick, D., and A. J. Keck, 1987: The importance of the lower level wind shear profile in tornado/non-tornado discrimination. Proc. Symp. Mesoscale Analysis and forecasting. Vancouver, British Columbia, Canada, ESA SP-282, 393-397.

Quiring, R. F., 1973: Low humidity: Still a problem with the modified NWS radiosonde? Bull. Amer. Meteor. Soc., 54, 551-552.

Rackliff, P. G., 1962: Application of an instability index to regional forecasting. Meteorol. Mag., 91, 113-120.

Rasmussen, E. N., 1994: VORTEX operations plan. National Severe Storms Laboratory, Norman, Oklahoma, 224 pp.

\_\_\_\_, and J. M. Straka, 1998: Variations of supercell morphology. Part I: Observations of the role of upper-level storm-relative flow. Mon. Wea. Rev., 126, 2406–2421.

\_\_\_\_, and D. O. Blanchard, 1998: A baseline climatology of sounding-derived supercell and tornado parameters. Wea. Forecasting, 13, 1148-1164.

Rodgers, D. M., D. L. Bartels, R. D. Menard, and J. H. Arns, 1984: Experiments in forecasting mesoscale convective weather systems. Preprints. Tenth Conf. Weather Forecasting and Analysis. Clearwater Beach, Amer. Meteor. Soc, 486-491.

Rotunno, R., and J. B. Klemp, 1982: The influence of the shear-induced pressure gradient on thunderstorm motion. Mon. Wea. Rev.,110, 136–151.

\_\_\_\_, and \_\_\_\_, 1985: On the rotation and propagation of simulated supercell storms. J. Atmos. Sci.,42, 271–292.

Schwartz, B. E. and C. A. Doswell III, 1991: North American rawinsonde observations: Problems, concerns, and a call to action. Bull. Amer. Met. Soc., 72, 1885-1896.

Schroeder, S. R., 2003: Completing instrument metadata and adjusting biases in the radiosonde record to allowdetermination of global precipitable water trends. Preprints, 12th Symp. On Meteorological Observations and Instrumentation, Long Beach, CA, 2003. .

Showalter, A. K., 1953: A stability index for

thunderstorm forecasting. Bull. Amer. Meteor. Soc.,34, 250–252. *This is given erroneously in some sources, including the AMS Glossary, as 1947.*

Stackpole, J. D., 1967: Numerical analysis of atmospheric soundings. J. Appl. Meteor., 6, 464-467.

Subramanian, D. V., and P. S. Jain, 1966: Stability index and area forecasting of thunderstorms. Preprints. 12th Conf. Radar Meteorology. Norman, Amer. Meteor. Soc, 156-159.

Thompson, R. L. and R. Edwards, 2000: An overview of environmental conditions and forecast implications of the 3 May 1999 tornado outbreak. Wea. Forecasting, 15, 682-699.

_____, R. Edwards and J.A. Hart, 2002: Evaluation and Interpretation of the Supercell Composite and Significant Tornado Parameters at the Storm Prediction Center. Preprints, 21st Conf. Severe Local Storms, San Antonio TX.

_____, _____, J. A. Hart, K. L. Elmore, and P. Markowski, 2003: Close proximity soundings within supercell environments obtained from the Rapid Update Cycle. Wea. Forecasting, 18, 1243-1261.

_____, R. Edwards, J. A. Hart, K. L. Elmore, and P. Markowski, 2003: Close proximity soundings within supercell environments obtained from the Rapid Update Cycle. Wea. Forecasting, 18, 1243-1261.

_____, C. M. Mead, and R. Edwards, 2004a: Effective bulk shear in supercell thunderstorm environments. Preprints, 22nd Conf. on Severe Local Storms, Hyannis, MA, Amer. Meteor. Soc.

_____, R. Edwards, and C. M. Mead, 2004b: An update to the Supercell Composite and Significant Tornado Parameters. Preprints, 22nd Conf. on Severe Local Storms, Hyannis, MA, Amer. Meteor. Soc.

_____, and C.M. Mead, 2006: Tornado Failure Modes in Central and Southern Great Plains Severe Thunderstorm Episodes. Preprints, 23rd Conf. Severe Local Storms, St. Louis MO.

_____, C. M. Mead, and R. Edwards, 2007: Effective storm-relative helicity and bulk shear in supercell thunderstorm environments. Wea. Forecasting, 22, 102–115.

_____, 2017. Personal communication, November 2017.

U. S. Weather Bureau, 1931: The Weather Bureau. U. S. Department of Agriculture, Misc. Publication No. 114, 34 pp.

Wade, C. G., 1994: An evaluation of problems affecting the measurement of low relative humidity on the United States radiosonde. J. Atmos. and Oceanic Tech., 11, 687-700.

Weisman, M. L., and J. B. Klemp, 1982: The dependence of numerically simulated convective storms on vertical wind shear and buoyancy. Mon. Wea. Rev.,110, 504–520.

_____, and _____, 1984: The structure and classification of numerically simulated convective storms in directionally varying wind shears. Mon. Wea. Rev.,112, 2479–2498.

_____, and _____, 1986. Characteristics of isolated convective storms. Mesoscale Meteorology and Forecasting. P. Ray, Ed., Amer. Meteor. Soc., ch. 15, . 504–520. BRN #2.

Wicker, L. J., 1996: The role of near surface wind shear on low-level mesocyclone generation and tornadoes. Preprints, 18th Conf. on Severe Local Storms, San Francisco, CA. Amer. Met. Soc. 115-119.

Wilhelmson, R. B., and J. B. Klemp, 1978: A numerical study of storm splitting that leads to long lived storms. J. Atmos. Sci., 35, 1974–1986.

World Meteorological Organization, 2017: The environmentally friendly radiosondes. CIMO/ET-A1-A2 Doc. 6, 4 pp.

Zeitler, J. W., and M. J. Bunkers, 2005: Operational forecasting of supercell motion: Review and case studies using multiple datasets. Natl. Wea. Dig., 29 (1), 81–97.

Zrnić, D.S., 1987: Three-body scattering produces precipitation signatures of special diagnostic value. Radio Sci., 22, 76-86.

# Index

# Index

## Symbols

0-1 km shear  124
0-1 km, SRH  126
0-3 km shear  124
30R75 method  109

## A

absolute humidity  28
absolute instability  74
ACARS  58
Adedokun Index  180
adiabatic cooling  23
advection  13
aerosol  24
air mass cooling  23
ALSTG  4
altimeter setting  4
altitude  1
  density  3
  geometric  2
  pressure  3
AMDAR  58
anvil-layer shear
  and SHARPpy  192

## B

balloon  40
balloons
  hydrogen  37
black body radiation  12
BRN  118, 162
BUFR  54
bulk Richardson number  118
Bulk Richardson number  162
bulk shear  107, 116
Bunkers method, storm movement  111
buoyancy  13

## C

cap  96, 136
capacitive
  hygrometer  51
capacitive hygrometer  51
capacitor
  thermal  44
CAPE  134
  normalized  90
capping inversion  96
carbon hygristor  51
case studies  131
CCL  81
CCN  25
centripetal acceleration  2
chilled mirror hygrometer  52
CIN  136
CINH  136
circulation
  solenoidal  96
cloud condensation nuclei  25
composite sounding  85
condensation  19
conditional instability  72
conduction  14
contamination  46
  of temperature  46
continental tropical  93
convection  13
convective condensation level  81
convective inhibition  136
convective instability  75
convective temperature  81
cooling
  adiabatic  23
  air mass  23
  evaporational  23
  isobaric  22
coordinates
  skew-T  66
Corfidi vectors  123
corrections
  wind  114
critical angle  121, 160
cross section
  RAOB  208
Cross Totals  154
CT  154

## D

DCAPE  148
deep-layer shear  116
density  7, 19
density altitude  3
deposition  17
depression  22
dewpoint
  and skew-T log-p  68
dewpoint temperature  26
differential instability  79
Downdraft CAPE  148
downwind propagation  123
dry adiabats  67
dryline  92
dust  24

## E

effective bulk shear  118
effective layer  90
effective layer, helicity  126
EHI  166
  and VGP  164
electromagnetism  9
emagram  63
endothermic process  16
energy-helicity index  166
equilibrium  21
equilibrium saturation point  21
equivalent potential temperature  19, 27, 82
  and SHARPpy  192
equivalent temperature  82
Espy
  James P.  36
ESRL  215
evaporation  16
evaporational cooling  23
evaporative cooling  27

exothermic process 16

## F

Fawbush-Miller Index 181
Fawbush-Miller method, hail 181
flight train 39
folds
  tropopause 29
forecast index 131
freezing 16
frontal inversion 72

## G

gas 15
geopotential 2
goldbeater's skin 48
GPS 53
gravity 2
ground-relative wind 104

## H

heat 8
helicity 124
  effective layer 126
  ID 126
  storm-relative 124
heterogeneous nucleation 24
hodograph 103, 131
  SHARPpy 189
  straight-line 108
homogeneous nucleation 24
humidity
  and radiosondes 48
hydrogen balloons 37
hygristor
  carbon 51
hygrometer
  chilled mirror 52
hysteresis 47

## I

ideal gas law 19
ID, helicity 126
inflow 120
  storm-relative 120, 121

inflow shear 120
infrared 11
  longwave 11
  shortwave 11
insolation 11
instability
  absolute 74
  conditional 72
  convective 75
  differential 79
  MAUL 77
  moist absolute 77
  potential 75
  symmetric 76
internal dynamics (ID) method 112
inversion 72
  frontal 72
  radiation 72
  subsidence 72
irradiance 11
ISA 69
isentropes 67
isobaric cooling 22
isobars 66
isotherms 66

## J

Jefferson Index 180

## K

KI 150
K Index 150
kites 35

## L

lapse rate
  superadiabatic 75
LCL 80, 140
left mover 113
level of free convection 82, 142
LFC 82, 142
LI 144
lid 136
Lid Strength Index 181

lift
  orographic 79
lifted condensation level 80, 140
lifted index 144
  forecast form 144
  static form 144
lifted parcel 86
light, and balloon 40
liquid 15
lithium chloride 51
longwave radiation 11
LORAN-C 53

## M

magnitude
  shear 107
maritime tropical 93
mass 7
MAUL 77
MCSs
  and Corfidi vectors 123
MDCRS 58
melting 16
meso-gamma 31
mesoscale 30
microscale 30
mini-supercells 138
mixed layer parcel 88
mixing ratio 25
  and skew-T log-p 68
moist absolute instability 77
moist adiabats 67
most unstable parcel 88
motion
  storm 108

## N

National Center for Atmospheric
  Research 212
National Profiler Network 56
NCAPE 90, 138
NCAR 212
nitrates 24
normalized CAPE 90, 138
nucleation 24
  heterogeneous 24

homogeneous 24

## O

OLR 12
Omega 53
orographic lift 79
outgoing longwave radiation 12

## P

parachute 40
parcel
  effective layer 90
  lift 86
  mixed layer 88
  most unstable 88
  surface based 86
phase change 14, 24
phlogiston 23
PIBAL 38
pilot balloon 38
PivotalWeather 214
Planck's Law 11
positive area 90
Possible Hazard Type 193
potential instability 75
potential temperature 18, 80
  and skew-T log-p 67
PPBB 54
precipitable water 28
pressure 4
  and skew-T log-p 66
  radiosonde 42
  saturated vapor 85
  saturation vapor 20
  sea level 5
  station 4
  water vapor 20
pressure altitude 3
profiler 56
propagation
  downwind 123
  upwind 123
proximity sounding 85

## Q

QFE 4
QFF 5
QNH 4

## R

Rackliff Index 180
radar
  WSR-88D 54
radiation 9
  black body 12
  outgoing longwave 12
radiational inversion 72
radiosonde 37, 39, 41
  models chart 217
  network 218
Radiosonde Replacement System 54
radiotheodolite 53
RAOB 205
  cross section 208
RAP 215
Rapid Refresh 215
relative humidity 27
  and skew-T log-p 69
right mover 113
RRS 54
RS92, Vaisala 43
RUC 215

## S

SARS 201
saturated equivalent potential temperature 83
saturated vapor pressure 85
saturation 21
saturation mixing ratio 26
saturation vapor pressure 20
scale 30
  storm scale 31
SCP 168
sea level pressure 5
sedimentation 19
SHARPpy
  equivalent potential temperature 192
  hodograph 189
  Possible Hazard Type 193
  significant tornado parameter 203
Sounding Analog Retrieval System 201
storm-relative wind 192
Storm Slinky 190
shear 107
  0-1 km 124
  0-3 km 124
  anvil-layer 192
  bulk 116
  bulk Richardson 118, 120
  deep-layer 116
  effective bulk 118
  inflow 120
  total 107
shear vector 107
SHIP 174
shock unit 40
shortwave radiation 11
Showalter index 146
SI 146
significant hail parameter 174
significant tornado parameter 170
  and SHARPpy 203
skew-T log-p
  development 63
Snow White 52
solenoidal circulation 96
solid 15
sounding
  composite 85
  proximity 85
  spaghetti 85
Sounding Analog Retrieval System 201
spaghetti sounding 85
SPC 213
specific humidity 28
specific volume 19
SRH 158
SSI 146
stability
  static 70, 71
stability index 131
standard atmosphere 69
static stability 70, 71
station pressure 4

storm motion 108
  30R75 109
  Bunkers method 112
  internal dynamics 112
  left mover 113
  right mover 113
Storm Prediction Center 213
storm-relative helicity 124, 158
  0-1 km 126
  effective layer 126
  ID 126
storm-relative inflow 120, 121
storm-relative motion 115
storm-relative wind
  and SHARPpy 192
storm scale 31
Storm Slinky 190
STP 170, 203
straight-line hodograph 108
stratosphere 29
Stüve diagram 63
sublimation 16
subsidence 19
subsidence inversion 72
sulfates 24
superadiabatic lapse rate 75
supercell composite parameter 168
supercells
  mini 138
supersaturation 22
surface based (SB) 86
SWEAT 172
symmetric instability 76
synoptic scale 30

## T

TEMP 54
temperature 6, 17
  and skew-T log-p 66, 68
  convective 81
  dewpoint 26
  equivalent 82
  equivalent potential 19, 27, 82
  parcel 17
  potential 18

radiosonde 43
saturated equivalent potential 83
virtual 18, 92
wet-bulb 26, 84
wet-bulb potential 84
tephigram 63
thermal capacitor 44
thermistor 44
theta 18
theta-e 19, 27
thickness
  and skew-T log-p 68
total energy index 180
total shear 107
Total Totals 156
train
  flight 39
tropopause 29
tropopause folds 29
troposphere 29
TTAA 54
TTBB 54
TTI 156

## U

UCAR 212
University Corporation for Atmospheric Research 212
University of Wyoming 211
upwind propagatiion 123

## V

vapor 15
vapor pressure 84
vector 104
  shear 107
vectors 104
velocity
  azimuth 55
  radar 55
Vertical Totals 70, 152
VGP 164
virtual temperature 18, 92
volume 7
  specific 19

vorticity generation parameter 164
VT 152

## W

water 19
water vapor 19
water vapor pressure 20
wavelength 9
WBZ 176
wet-bulb potential temperature 84
wet bulb temperature 26, 84
wet-bulb zero height 176
wind
  and radiosonde 53
  ground-relative 104
wind profiler 56
WSR-88D 54

CPSIA information can be obtained
at www.ICGtesting.com
Printed in the USA
FSHW010150311218
54521FS